Managing Data as a Product

Design and build data-product-centered
socio-technical architectures

Andrea Gioia

Managing Data as a Product

The author acknowledges the use of cutting-edge AI, such as ChatGPT, with the sole aim of enhancing the language and clarity within the book, thereby ensuring a smooth reading experience for readers. It's important to note that the content itself has been crafted by the author and edited by a professional publishing team.

Group Product Manager: Apeksha Shetty
Publishing Product Manager: Nilesh Kowadkar
Book Project Manager: Aparna Nair
Senior Content Development Editor: Priyanka Soam
Technical Editor: Sweety Pagaria
Copy Editor: Safis Editing
Proofreader: Priyanka Soam
Indexer: Rekha Nair
Production Designer: Joshua Misquitta
Senior DevRel Marketing Executive: Nivedita Singh

First published: November 2024
Production reference: 2291124

Published by Packt Publishing Ltd.
Grosvenor House
11 St Paul's Square
Birmingham
B3 1RB, UK.

ISBN 978-1-83546-853-1
www.packtpub.com

To my parents, Delia and Salvatore, for your endless encouragement, patience, and love. This is for you.

– Andrea Gioia

Foreword

I have known and worked with Andrea for more than 7 years. He is undoubtedly recognized as an innovation leader in the field of data management and one of the most active thinkers in the global data community.

In my professional career, Andrea has been, and still is, a mentor, an example to follow for the passion he transmits in the continuous search for advancements in all aspects of data management.

We live in the era of the Fourth Industrial Revolution: in this scenario of uncertainty, companies must be able to continuously build new, dynamic capabilities to obtain competitive advantage and survive. From this point of view, core enterprise data is an essential resource that companies can leverage; however, data is a liquid asset and only generates value if it is actually used and, even more so, reused. Storing data without using it does not generate value, but only liability. To encourage the reuse of data and derive value from it, it is necessary to view data no longer as a by-product of software applications but as a first-class product in the enterprise architecture.

In this volume, Andrea highlights that managing data as a product is not a purely technological issue but requires an organic understanding of the company's business strategy and operations at 360 degrees, considering the market context, the business architecture, the organizational architecture, the operating model, the state of the IT architecture and many other aspects. In the current post-digitalization era, a data management practitioner needs more and more T-shaped transversal skills to successfully implement the change of perspective that leads to managing data as a product.

Andrea has been an example of a complete professional for years, with an open mind, always aimed at considering all these aspects. Through his advisory experience alongside customers, he has been able to understand that modularity in architecture design is an essential feature to manage the intrinsic complexity of a modern data ecosystem. This belief led him to define the paradigm described in this book for the development of modern data architectures.

However, as a profound expert of modern data management paradigms, by which he was inspired for the birth of this book, Andrea understood that the modularity of architecture alone is not enough to guarantee effectiveness in the use of data assets if not supported by a shared enterprise knowledge model. He realized that data management must go beyond the pure management of elementary data and extend to organically manage multiple types of metadata and corporate knowledge, focusing on the explicit modeling of domain semantics. This is a human-centric process that can boost the accuracy of new generative AI techniques applied to data but requires the active contribution of domain experts. If this company-specific, non-replaceable knowledge base is evolved in synergy with the development of data as products, it's easier to enable the actual composability of the data products that make the modular architecture and encourage their reuse, maximizing overall value for the company.

This book disseminates the main elements of the vision that the author has developed over the years, enriched with the fruit of the learnings that he has collected during his journey in advisory. It is a piece of work that will inspire many people in the evolution of their professional lives – not only data engineers but also data managers, chief data officers, chief information officers, and enterprise architects.

Finally, I would like to thank Andrea for asking me to be a reviewer of this volume. It was a privilege for me!

Giulio Scotti
Governance Advisor, Quantyca

Contributors

About the author

Andrea Gioia is a partner and CTO at Quantyca, a consulting firm specializing in data management, and co-founder of blindata.io, a SaaS platform for data governance and compliance. With over 20 years of experience, Andrea has led cross-functional teams delivering complex data projects across multiple industries. As CTO, he advises clients on defining and executing their data strategies. Andrea is a frequent speaker and writer, serving as the main organizer of the Data Engineering Italian Meetup and leading the Open Data Mesh Initiative. He is an active DAMA member and has been part of the DAMA Italy Chapter's scientific committee since 2023.

This book would not have been possible without the collaboration and support of everyone I work with daily to drive innovation in the data management market. I am deeply grateful to all my colleagues at Quantyca and Blindata, as well as to our partners and clients.

About the reviewer

Andrew Jones is an independent data consultant, helping data leaders transform their organizations into ones where data drives revenue - is dependable, governed, and valuable, and is guaranteed with data contracts. In 2021, he created data contracts, an architectural pattern that brings data teams closer to the business, drives team autonomy, and embeds quality and governance as part of the data platform.

In 2023, he wrote the definitive book on data contracts, *Driving Data Quality with Data Contracts*. Through his independent consulting practice, he helps organizations big and small adopt data products, data meshes, and data contracts. He is based near London, UK, and is a regular writer and public speaker.

Table of Contents

Part 1: Data Products and the Power of Modular Architectures

1

2

3

Data Product-Centered Architectures   53

Part 2: Managing the Data Product Lifecycle

4

Identifying Data Products and Prioritizing Developments   85

5

Designing and Implementing Data Products 115

6

Operating Data Products in Production 145

Part 3: Designing a Successful Data Product Strategy

8

9

Team Topologies and Data Ownership at Scale   227

10

Distributed Data Modeling   253

11

Building an AI-Ready Information Architecture   283

12

Bringing It All Together   313

Preface

Hello, and welcome to *Managing Data as a Product*! I'm excited to share everything I've learned about managing data as a product and how this new paradigm can solve recurrent problems in data architectures that, despite huge investment, periodically collapse under the weight of their own complexity, making sustainable evolution a real challenge.

Ironically, the most successful data platforms, those that bring the greatest value to an organization, are often the first to struggle. Their success drives rapid growth in both the number of managed data assets and users, which leads to complexity. This complexity gradually slows down their growth until the platforms become too costly to maintain and too slow to evolve. However, this march toward self-destruction isn't inevitable. We can rethink how we design data management solutions, so they don't fall victim to their success but instead exploit it, multiplying the value they generate for the organization while growing.

Managing data as a product allows us to handle growing complexity by modularizing the data management architecture. Each data product is a modular unit that helps isolate complexity into smaller, manageable parts. Over time, the collection of developed data products forms a portfolio of building blocks that can be easily recombined to support new use cases. This way, while the platform's complexity remains stable as it grows, the value derived from the managed data assets increases. Implementing new business cases becomes simpler, as existing data products can be reused rather than creating new ones from scratch.

However, managing data as a product is a profound paradigm shift from traditional monolithic data architectures, impacting not only technology but also, and especially, the organization. Throughout this book, chapter by chapter, we'll explore practical, actionable steps to adopt this new paradigm, addressing all key aspects from both a technical and organizational perspective.

As we'll see, adopting a data-as-a-product approach is challenging, but it's well worth the effort. This book is a travel guide inspired by my experience, aimed at helping you find the best path for your unique context to successfully navigate this paradigm shift.

What this book covers

Chapter 1, From Data as a Byproduct to Data as a Product, shows how modularizing data architecture with data products solves recurring problems that make its sustainable evolution challenging over time.

Chapter 2, Data Products, defines what a data product is, outlining its key characteristics and explaining the essential components that make it up, highlighting how each element contributes to its overall function and value.

Chapter 3, Data Product-Centered Architectures, explores the foundational principles of a data product-centered architecture, analyzing the key operational and organizational capabilities required to manage it. We also compare other modern approaches such as data meshes and data fabrics with the data-as-product paradigm to highlight their similarities and key differences.

Chapter 4, Identifying Data Products and Prioritizing Developments, explains how to identify and prioritize data products using a value-driven approach. It starts by identifying relevant business cases through domain-driven design and event storming, then shows how to define the data products needed to support those business cases.

Chapter 5, Designing and Implementing Data Products, explores the process of designing a data product based on identified requirements, starting with techniques for defining scope, interfaces, and ecosystem relationships. It then examines the core components of a data product, their development process, and how to describe them with machine-readable documents. Finally, it analyzes the data flow, focusing on components responsible for sourcing, processing, and serving data.

Chapter 6, Operating Data Products in Production, covers the entire lifecycle of a data product, from release to decommissioning. It introduces CI/CD methodologies, explores managing a data product in production with a focus on governance, observability, and access control, and discusses techniques for evolving and reusing data products in a distributed environment.

Chapter 7, Automating Data Product Lifecycle Management, explains how to speed up the adoption of a data product-centric paradigm by creating a self-serve platform to mobilize the entire data ecosystem. It covers the platform's main features, how it improves the experience for developers, operators, and consumers, and the key factors in deciding whether to build, buy, or use a hybrid approach in implementing it.

Chapter 8, Moving through the Adoption Journey, covers the adoption of the data-as-a-product paradigm. It outlines the key phases of the process, exploring objectives, challenges, and activities for each stage. Finally, it discusses how to create a flexible data strategy that evolves with each phase, building on previous learnings.

Chapter 9, Team Topologies and Data Ownership at Scale, explains how to design an organizational structure for managing data as a product. It introduces the team topologies framework, including team types and interaction modes, and explores how to organize teams for efficient data product delivery. Finally, it looks at how to integrate these teams into the organization and decide between the centralized or decentralized data management model.

Chapter 10, Distributed Data Modeling, examines data modeling in a decentralized, data product-centered architecture. It defines data models and emphasizes intentionality in modeling, then examines physical modeling techniques for distributed environments. Finally, it covers conceptual data modeling and its role in guiding the design and evolution of data products within a cohesive ecosystem.

Chapter 11, *Building an AI-Ready Information Architecture*, explores how to build an information architecture that maximizes the value of managed data, starting with developed data products. It covers how different planes of the information architecture add context to data and focuses especially on the knowledge plane, where shared conceptual models ensure semantic interoperability between data products. Finally, it explores how federated modeling teams can create and link conceptual models to physical data, forming an enterprise knowledge graph crucial for unlocking the potential of generative AI.

Chapter 12, *Bringing It All Together*, revisits key concepts from earlier chapters, tying them to the core beliefs about data management that inspired this book. It wraps up with practical advice for becoming a more successful data management practitioner.

To get the most out of this book

In this book, both data products and the self-serve platform needed to support their development and operation are described at a logical level, without reference to any specific technology stack. Therefore, no prior knowledge of specific technologies is required to read and understand the content.

In some chapters, examples of metadata are provided to describe the components of a data product. This metadata is generally represented as JSON snippets. To use and modify them, we suggest a text editor that can recognize JSON syntax, such as Visual Studio Code.

If you are using the digital version of this book, we advise you to type the code yourself or access the code from the book's GitHub repository (a link is available in the next section). Doing so will help you avoid any potential errors related to the copying and pasting of code.

Download the example code files

You can download the example code files for this book from GitHub at `https://github.com/PacktPublishing/Managing-Data-as-a-Product`. If there's an update to the code, it will be updated in the GitHub repository.

We also have other code bundles from our rich catalog of books and videos available at `https://github.com/PacktPublishing/`. Check them out!

Conventions used

There are a number of text conventions used throughout this book.

`Code in text`: Indicates code words in text, database table names, folder names, filenames, file extensions, pathnames, dummy URLs, user input, and Twitter handles. Here is an example: "The `Promises` object contains all the metadata through which the data product declares the intent of the port."

A block of code is set as follows:

```
:purchases a rdf:Property ;
    rdfs:domain :Customer ;
    rdfs:range [ a owl:Class ;
        owl:unionOf ( :Product :Service ) ]
```

Bold: Indicates a new term, an important word, or words that you see onscreen. For instance, words in menus or dialog boxes appear in **bold**. Here is an example: "Select **System info** from the **Administration** panel."

> **Tips or important notes**
> Appear like this.

Get in touch

Feedback from our readers is always welcome.

General feedback: If you have questions about any aspect of this book, email us at customercare@packtpub.com and mention the book title in the subject of your message.

Errata: Although we have taken every care to ensure the accuracy of our content, mistakes do happen. If you have found a mistake in this book, we would be grateful if you would report this to us. Please visit www.packtpub.com/support/errata and fill in the form.

Piracy: If you come across any illegal copies of our works in any form on the internet, we would be grateful if you would provide us with the location address or website name. Please contact us at copyright@packt.com with a link to the material.

If you are interested in becoming an author: If there is a topic that you have expertise in and you are interested in either writing or contributing to a book, please visit authors.packtpub.com.

Share your thoughts

Once you've read *Managing Data as a Product*, we'd love to hear your thoughts! Scan the QR code below to go straight to the Amazon review page for this book and share your feedback.

https://packt.link/r/1-835-46853-5

Your review is important to us and the tech community and will help us make sure we're delivering excellent quality content.

Free Benefits with Your Book

This book comes with free benefits to support your learning. Activate them now for instant access (see the "*How to Unlock*" section for instructions).

Here's a quick overview of what you can instantly unlock with your purchase:

PDF and ePub Copies　　　　　**Next-Gen Web-Based Reader**

Access a DRM-free PDF copy of this book to read anywhere, on any device.

Use a DRM-free ePub version with your favorite e-reader.

Multi-device progress sync: Pick up where you left off, on any device.

Highlighting and notetaking: Capture ideas and turn reading into lasting knowledge.

Bookmarking: Save and revisit key sections whenever you need them.

Dark mode: Reduce eye strain by switching to dark or sepia themes

How to Unlock

Scan the QR code (or go to `packtpub.com/unlock`). Search for this book by name, confirm the edition, and then follow the steps on the page.

Note: Keep your invoice handy. Purchases made directly from Packt don't require one

Part 1:
Data Products and the
Power of Modular Architectures

In this part, we'll explore the rationale and implications of treating data as a product. We'll break down the key components of a data product, starting with interfaces accessible to external consumers, moving through internal data management applications, and ending with the infrastructure needed for production use. Finally, we'll highlight the qualities and capabilities that a data architecture centered on data products must embody to balance the agility needed for scalability with the governance essential for sustainability.

This part has the following chapters:

- *Chapter 1, From Data as a Byproduct to Data as a Product*

- *Chapter 2, Data Products*

- *Chapter 3, Data Product-Centered Architectures*

1
From Data as a Byproduct to Data as a Product

In this book, we will explore how to transition from managing data merely as a byproduct that supports applications to managing data as a product in its own right. Before tackling the various aspects that contribute to this paradigm shift, it's crucial to understand why managing data as a product is important and how this practice enables us to surpass the limits of today's data platforms.

In this chapter, we will explore the history of monolithic data platforms, which have characterized the evolution of data management over the last 30 years. We will seek to understand the common problems that make them incapable of sustainably managing the accidental complexity they generate as they grow. Finally, we will see why addressing the fundamental issues, instead of merely treating surface-level symptoms, requires more than just technological innovations. It calls for a paradigm shift that leads us toward more sustainable socio-technical architectures, based on the key practice of managing data as a product.

This chapter will cover the following main topics:

- Reviewing the history of monolithic data platforms
- Understanding why monolithic data platforms fail
- Exploring why we need to manage data as a product

Reviewing the history of monolithic data platforms

Managing the substantial amount of data that's generated by every company daily is a complex endeavor. It calls for dedicated resources and technological support in the form of a specific data platform.

Nowadays, data platforms often fall short in delivering the expected value compared to the investments made, primarily due to organizations' inability to sustainably manage the complexity they generate over time.

> **System complexity**
>
> The complexity of a system is determined by the number of its components multiplied by the number of correlations between them. A database with 10,000 tables is not much more complex than a database with 100 tables if the tables themselves are not correlated. Each table tells its own story. It can be manipulated without concern for the meaning of other tables and the potential impacts that the executed action may have on them. However, the complexity between the two databases is very different if the tables are highly correlated. In systems with high levels of correlation, complexity generally grows quadratically with the number of components.

Not all the complexity of a system is related to the complexity of the specific function for which the system is designed (**essential complexity**). Part of the complexity is also derived from the approaches and tools used to develop it (**accidental complexity**). Essential complexity is incompressible, but accidental complexity can be reduced or at least kept under control to prevent it from growing to the point where the entire system becomes unmanageable and incapable of evolving.

To create systems capable of supporting the complexity they generate, it is necessary to intervene in tools and development approaches that generate accidental complexity. Rarely in my experience have I seen a data platform fail due to a wrong technological choice (that is, tools). Even when it did happen, the blame could not be squarely placed on individual tools but on the incorrect manner of using them within a coherent and purposeful architecture. If data platforms fail, crushed under the weight of their complexity, it is predominantly due to the socio-technical architecture approaches that were employed to develop and evolve them over time (that is, approaches).

For this reason, modern approaches to data management focus on these architectural aspects rather than individual technology tools. However, it has not always been this way. The subsequent approaches to data management that we have seen over the years have been primarily driven by technologies, seen as key to solving data management issues. Before delving into the common elements leading to the failure of monolithic data architectures in more detail, let's explore the main approaches that have evolved over time and on which most monolithic platforms are based.

Data warehouse

The concept of a **data warehouse** (**DWH**) was first introduced in the late 80s by IBM researchers Barry Devlin and Paul Murphy. They proposed the idea of a centralized repository to store and manage large volumes of data from various sources to support decision-making processes.

The adoption of DWH began to spread widely in the 90s, thanks to the conceptual work of authors such as Bill Inmon and Ralph Kimball. Before that, data integrations were predominantly carried out using point-to-point logic, while analyses were mostly performed by directly extracting data from source systems.

This tactical approach to data management made perfect sense in the early days of the third industrial revolution when the digitalization of business processes was just beginning. The applications were still few, and the produced data was limited. The focus of IT strategy was rightly on automating core business processes through new applications, rather than managing the data assets they made available.

However, as digitalization progressed, the number of applications and generated data quickly reached a level where managing integrations with point-to-point logic became too complex and no longer sustainable. The DWH emerged as a response to these scalability issues, introducing the idea of having a dedicated platform and team for data management. The DWH represents a significant shift in both technological and organizational architecture. With the introduction of the DWH, data management becomes a full-fledged organizational function with dedicated resources, processes, and objectives.

From a technological architecture viewpoint, the DWH is based on the idea of collecting all data generated by digitalized processes in a single repository, consolidating them, and making them available for analysis through a unified model. The DWH assumes the role of a single **source of truth** for company data, serving as the foundation for all analyses and decision-making processes. It also decouples data producers from consumers, allowing the costs of integration and consolidation for a specific data domain to be incurred only once and then reused for the implementation of multiple analytical use cases.

However, like any architecture, the DWH has its limitations. Specifically, it prioritizes the quality and reusability of consolidated data in the unified model over agility in development. Consequently, while integration costs decrease, and data consumption is simplified on one hand, on the other hand, the lead time for integrating new data sources and releasing new analyses increases.

The end of the millennium witnessed a constant growth in the number of applications within organizations, leading to a tremendous increase in the volume, variety, and velocity of data to manage. This shift resulted in a transition from simply talking about data to addressing the new challenges of **big data**. Simultaneously, the number and variety of data consumers for analytical purposes significantly grew. Traditional directional reporting expanded to include analyses for a broader audience, **machine learning (ML)**-based models, and various data-driven applications.

Due to these trends, at the beginning of the millennium, the advantages of DWHs began to be overshadowed by their disadvantages. The DWH became a bottleneck, squeezed between data producers and consumers, increasingly seen not as an enabler but as a hindrance to digital transformation initiatives.

Moreover, the growth in the volume, variety, and velocity of data, along with diverse consumption patterns, made DWHs unsuitable for optimally managing all types of data and workloads. Finally, the surge in data volume led to a spike in the costs of the hardware and software that was used to implement DWH platforms – costs that were increasingly less justified by the **return on investment (ROI)** of analytical initiatives.

Data lake

The **data lake** approach began gaining traction in the early 2000s as a response to the challenges faced by the DWH in managing big data and new analytical workloads, particularly those related to ML and **artificial intelligence (AI)**. From a logical architecture perspective, a data lake, like a DWH, serves as a centralized repository of data. However, from a technological architecture standpoint, it relies on distributed storage and computing solutions, such as **HDFS** and **MapReduce**, among the first available on the market. These distributed technologies are built on commodity hardware and open source software, making the costs for managing big data volumes significantly lower than those associated with traditional analytical databases used by DWHs.

The adoption of this approach in the early 2000s was not solely driven by cost considerations. In general, vendors entering this market positioned data lakes as platforms for more agile data management compared to traditional DWHs. Specifically, the separation between storage and computation allows for direct querying of raw data (**schema on read**), theoretically eliminating the need for extensive upfront processing to make data available to downstream consumers.

The combination of cost savings and the promise of agility led to the rapid uptake of this approach, even in many companies that didn't necessarily have truly big data to manage.

The initial idea of addressing the need for agility in developing new analytical solutions by focusing almost exclusively on data collection, with minimal interventions in terms of structuring, cleaning, normalizing, and enriching the data, did not prove to be highly effective. Many early data lake projects that heavily emphasized this agile and destructured approach resulted in systems that were complex to manage and difficult to use, where data fragmentation from source systems was quickly reproduced within the data lake, turning it into a **data swamp**.

Today, modern data lakes strike a more balanced approach between the need for agility and the need for control. They generally rely on **medallion architecture**, a layered architecture where raw data is transformed and enriched progressively as it flows through different layers before being made available for consumption.

Data lakes have been a key element in the evolution of data management technologies, with significant impacts also on DWHs. The most notable impacts are as follows:

- Horizontal scaling on commodity hardware to effectively manage large quantities of data
- Separation between storage and computation and the corresponding ability to reuse the same data for different workloads
- Utilization of enterprise-level open source software supported by vendors through diverse commercial models, promoting a reduction in lock-in and accelerating innovation in the industry
- Reduction of costs for implementing analytics solutions alongside market democratization

DWH versus data lake

The data lake approach has not replaced the DWH and vice versa. Both have evolved, addressing their respective limitations.

Modern DWHs are built on distributed, horizontally scalable architectures. Storage and computation are separated and independently scalable. They have become multimodal data management platforms capable of handling various workloads on stored data. Additionally, they enable the management of diverse data types: structured, semi-structured, and even non-structured.

On the other hand, modern data lakes have progressively acquired functionalities typical of a DWH, such as transactionality, time travel, and SQL-based query interfaces. Query engines have now achieved performance levels comparable to those of DWHs. Many aspects related to data management optimization are automatically handled by the platform, broadening the user base and reducing barriers in terms of technical knowledge required for implementation.

The dichotomy that once existed no longer holds. The technologies that support both approaches can effectively handle all types of data across major workloads in a highly scalable manner. Even the approaches to data modeling and governance have converged. Modern data lake management approaches have become more organized and structured, while those for DWH management have become more flexible and faster.

It is possible to build a data platform based solely on one of these approaches and their respective technologies without worrying about potential limitations. However, many organizations still prefer a hybrid approach, leveraging the strengths of data lake technologies for flexibility and scalability in managing raw data, and the strengths of DWH technologies for achieving the best performance in analyzing enriched data.

Modern data stack

In the early 2010s, major cloud providers released their DWH in a fully managed mode. Simultaneously, they introduced fully managed versions of data lake platforms. In the same period, pure-player vendors also entered the market, offering cloud solutions for DWHs (for example, Snowflake) and the data lake (for example, Databricks). Within a short time, the market was flooded with solutions capable of ensuring the necessary scalability to handle large amounts of data in an elastic and fully managed manner. Traditional vendors initiated a process of porting their products to the cloud to avoid falling behind and becoming legacy. The momentary lack of cloud-native solutions for managing data on new cloud-based platforms led to the emergence of many startups. These covered the essential functionalities constituting the technological stack needed to manage the data life cycle in the cloud, from acquisition to analysis.

The second half of the 2010s witnessed a race between traditional players committed to transitioning their suites to the cloud and new players eager to gain as much market share as possible in specific functionalities. Taking advantage of the transition period to the cloud for traditional vendors and then focusing intensely on specific functionalities, the new players managed to capture significant market shares among organizations determined to migrate their analytical workloads to the cloud. This marked the beginning of a technological unbundling cycle characterized by an explosion of point solutions, a landscape that we still find ourselves in today. The **modern data stack** (**MDS**) is the term that's used to refer to this new ecosystem of specific and cloud-native solutions supporting data management.

The MDS has brought considerable technological innovation, further reducing accidental complexity associated with tools and consequently increasing the productivity of data teams. However, the extreme fragmentation of tools required for data management has increased the complexity of developing and operationally managing the core capabilities of a data platform. Today, data platform architectures are based on many more tools than in the past. These tools must be selected, integrated, and governed from both security and cost perspectives. In summary, while the innovative drive of the MDS ecosystem has reduced development times and analysis maintenance costs, on the other hand, it has increased operational costs in terms of developing and maintaining the underlying platform. It is not always clear whether the balance between advantages and disadvantages is positive or negative. It is likely that in the coming years, after a strongly expansive phase (unbundling), the offering will converge again toward a rationalization phase (bundling), where we may see some MDS vendors merging, others being acquired by big tech, and some potentially failing after the driving force of the collected investments diminishes.

The MDS is not a completely new approach to data management. As we have seen, it is a new technological proposition that integrates with the popular approaches mentioned previously.

Understanding why monolithic data platforms fail

If we look at the evolution of data management over the last 40 years, we'll see a story of incredible technological revolutions and just as many project failures. At the beginning of this chapter, we mentioned that the main reason for these failures is the complexity generated by data management platforms, and this complexity grows approximately quadratically with the size of the platform. Therefore, these are not typical project failures as we are accustomed to understanding them. Data platforms rarely fail before their launch, never making it into production. Instead, they often experience failures related to their ability to evolve and survive over time. Platforms don't fail immediately but over time, as they struggle to deliver the expected value in proportion to the constantly increasing maintenance costs they generate.

Like a Jenga tower becoming increasingly unstable as more pieces are added until it collapses, data platforms often implode under the weight of their own complexity (**complexity catastrophes**). At this point, you must start over and build a new platform from scratch. Many organizations have gone through different generations of data platforms throughout their history, trying new technological stacks offered by a constantly evolving market, but facing the same problems again and again.

Technology has helped solve some symptoms of the problems but not the root causes of accidental complexity. If not managed, this complexity inevitably brings us back to the starting point. These causes can be traced back to the monolithic architecture, both from a technological and organizational perspective, common to all data management approaches presented in the previous section. But what exactly does it mean for a data platform to have a monolithic architecture? Let's explore that together.

Monolithic versus modular architecture

A data platform, when it is first created, is generally not a complex system. It becomes complex as it grows because its components increase and the interconnections between them multiply. In nature, complex systems are either governed or self-governed by structuring their components into subsystems, organized in hierarchies or more complex topologies. By doing so, they manage to redistribute the total complexity of the system among its parts and govern it. Similarly, a data platform is not monolithic or less monolithic based on its level of socio-technical centralization. A data platform is considered monolithic when it cannot reorganize its socio-technical components into modules that act as distinct, autonomous subsystems, each capable of masking part of the system's overall complexity.

It's important not to confuse the concept of modules with the concepts of architectural components. Generally, monolithic platforms have (at least on slides) a clear architecture broken down into layers, with macro components for each layer. However, they are not modular. Modules are units of a larger system, structurally independent from each other but capable of working together. A modular system must be organized in an architecture that allows modules to maintain their structural independence on one hand but cooperate on the other. A module is a component of a system, but it's not necessarily true the other way around. Similarly, a modular system must have a clear architecture to ensure cooperation between modules, but it's not necessarily true that a system with a clear architecture is modular.

It's also crucial to observe that for a module to be considered as such and to preserve its structural integrity over time, it must not only have a clear interface that distinguishes its internal components from the rest of the system but must also ensure that its internal components are strongly connected and relatively loosely coupled to the other modules of the system. Some modules will be more connected than others, but overall, a modular system ensures decoupling between its modules. Therefore, it's not enough to draw circles around the components of a system to make the system modular. To become a module, components must be grouped based on their structural coherence.

The lack of modularity, not the lack of decentralization, as is often believed, is the main characteristic of a monolithic platform. Both monolithic and modular platforms can be more or less centralized at the technological and organizational levels without fundamentally changing their nature. Even a distributed platform, composed of components with a clear interface, can lack modularity if the level of coupling between modules is so high that it erodes the autonomy of each module being a **distributed monolith**.

So, we can define a monolithic platform as a non-modular platform. I introduced the concept of modularity as a tool to govern the accidental complexity generated by the growth of the platform, suggesting accordingly that the lack of modularity is one of the main structural problems leading to the failure of monolithic platforms. Let's learn how modularization can help us build platforms that do not reproduce the problems of the past.

The power of modularity

Modularization – in other words, the process of articulating a complex system into subsystems – is a practice that we constantly see applied in nature. The human body, for example, is an emblem of modularization, being organized into subsystems that work together at different levels of abstraction to ensure the correct functioning of the entire organism.

The benefits of modularization are evident not only in nature but also in many complex human activities. Since humans have limited cognitive capacities, to manage a complex system, they must divide it into smaller and sufficiently decoupled parts that can be analyzed separately. This modular structure allows humans to encapsulate the complexity of the system into functional parts defined by a clear interface. The interface, in addition to delimiting the perimeter of the module, provides an abstraction that allows part of the complexity to be hidden from the external observer (**information hiding**). It becomes possible to reduce **cognitive load** by deciding whether to reason about the internal functioning of a module, ignoring the rest of the system, or to reason about the interactions between the modules of the system, ignoring their internal functioning.

Modules can also be structured in a hierarchy with successive levels of abstraction. Modules at one level are composed of modules from the level below and define more abstract functions or concepts. For example, the human body is composed of various types of cells, cells can be grouped into organs, and organs can be grouped to form organ systems such as the respiratory system and the circulatory system. The structuring of modules in a modular system is another key element that can reduce cognitive load. It allows reasoning to be shifted across different levels of abstraction according to objectives, ignoring the details of the functioning of modules present in other levels.

The structure of modular systems can be regulated and evolved based on human cognitive capabilities. When a module becomes too complex, it can be divided into smaller modules. When modules become too numerous, they can be regrouped into macro modules belonging to a higher level of abstraction in the hierarchy.

Failure loops

The principles of encapsulation, information hiding, and abstraction, typical of a modular structure, are strategies that humans have always used, borrowing them from nature, to overcome our cognitive limitations that would not allow us to successfully manage activities or systems that are too complex.

These principles are systematically employed in many engineering practices, including software engineering. So, why do data platforms traditionally have a monolithic structure and not a modular one? Let's try to understand this together using **systems thinking**, which provides an excellent conceptual framework for analyzing the structure, interactions, and dynamics of a complex system.

In a system, components are interconnected. There are interactions between components, meaning that when one component acts on another, it changes its state and, consequently, its behavior. The component performing the action can, in turn, receive feedback following the executed action, modifying its state and, consequently, future behavior. Feedback can be direct when it comes without intermediaries from the component on which the action was performed, or indirect when it comes from another component following a more elaborate cycle of feedback. **Feedback loops** are the main tool used in systems thinking to holistically study the emergent behaviors of a system based on the interactions between its components.

There are two main types of feedback loops: **reinforcing loops** and **balancing loops**. Reinforcing loops amplify a specific behavior of the system while balancing loops reduce it. Therefore, feedback loops are control mechanisms that regulate the overall functioning of a system. For a system to survive, it must structure the interactions between its components in such a way that the resulting composition of these feedback loops does not make it too rigid and incapable of adapting but, at the same time, does not make it unstable and out of control. Systems are constantly seeking a balance between stability and the ability to change to survive. This equilibrium isn't static; rather, it's dynamic, with continual changes unfolding within the system, maintaining overall stability.

Unlike the ways we are accustomed to modeling reality, in complex systems, cause and effect and action and reaction are separated in time and space. They are separated in space because the action of one component can trigger feedback loops that create repercussions on components even very distant within the connection space. They are separated in time because an action on one component can lead to consequences on another component after some time; feedback can be delayed and asynchronous to the action. These nonlinear dynamics, orthogonal to our classical way of thinking, often lead to errors when analyzing the causes that lead to the emergence of certain behaviors in a system. System thinking seeks to provide more aware modeling methods of the actual functioning of a system to avoid these types of analysis errors. Let's look at a generic example of modeling a system based on the building blocks provided by systems thinking described so far before applying them to the systemic analysis of data platforms:

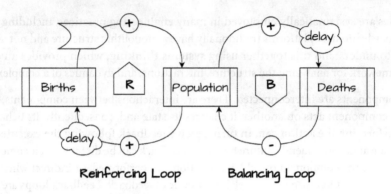

Figure 1.1 – Causal loop diagram

In the **causal loop diagram** (**CLD**) shown in *Figure 1.1*, relationships among different components of a system (population, births, and deaths) are depicted. Between births and population, there is a **reinforcing loop** – the more births there are, the larger the population becomes, and the larger the population, the more births there will be over time. Similarly, between the population and deaths, there is a **balancing loop** – the larger the population, the more deaths there will be over time, and the more deaths, the smaller the population becomes. Any action, whether from internal or external components of the system, causing an increase or decrease in births or deaths, leads to variations in the population. The population can change over time, growing or decreasing, so long as there is always a dynamic balance within the system that prevents it from going out of control (that is, extinction or unsustainable overpopulation).

The population example is based on a feedback loop structure that's common in many systems. This is known as the **balancing process**. Systems thinking has identified many of these recurring patterns, recognizable in various types of systems. To explain what makes modularization of data platforms challenging, we'll turn to two of these archetypes: shifting the burden and limits to growth.

Shifting the burden is a pattern that occurs when a quick fix is used to solve a problem, temporarily alleviating the symptoms without addressing the underlying issue. This makeshift solution is faster in relieving the symptoms of the problem and is therefore preferred over the actual cure, which requires more time and effort. Eventually, the problem reemerges, and the makeshift solution becomes the habitual response to deal with it. In the data platform context, this archetype happens when technology is used to mitigate all emerging problems, without focusing on the root causes. Several times, we believed we had found the ultimate solution to problems in a new technology, only to inevitably reproduce the same problems we hoped to solve each time. Over time, we have become overly dependent on technology, transforming data management into data technology management in recent years.

A CLD of the *shifting the burden* archetype, when applied to data platforms, is shown in *Figure 1.2*:

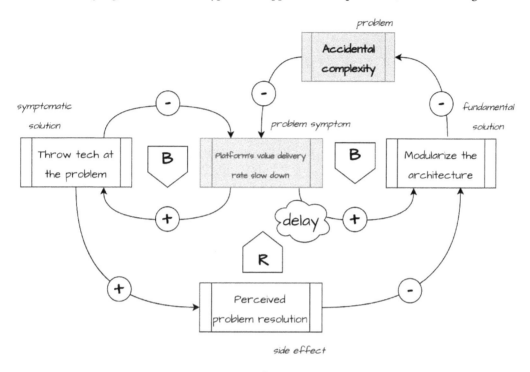

Figure 1.2 – Shifting the burden

Limits to growth is a pattern that manifests when a system has a limit to its growth. The system grows until it reaches the limit, at which point the actions that led to the growth start yielding diminishing returns. The cost to maintain the same levels of past growth becomes unsustainable, jeopardizing the stability of the system.

Within a data platform, the resource limiting its growth is the cognitive load that the data team can bear. Initially, it's not a problem. The platform is small, and the speed of integrating new data sources and developing new analyses is high. However, as the platform grows, so does its complexity. As mentioned earlier, in a data platform, complexity can increase rapidly with its size due to the high number of connections between data. Since the value of the platform lies more in the connections between data than in the data itself, this is the essential complexity inherent in its development. With the increase in complexity, the cognitive load on the development team also increases, and they would naturally be forced to slow down the delivery speed. As organizational pressures are generally high, instead of slowing down delivery, the team starts reducing the quality of developments, increasingly resorting to workarounds that introduce **technical debt**. Unmanaged technical debt leads to unorganized and often unnecessary growth of the platform, increasing its complexity – the accidental complexity in this case. Complexity, in turn, increases the cognitive load, triggering the reinforcing feedback loop R2, shown by the dashed line in *Figure 1.3*:

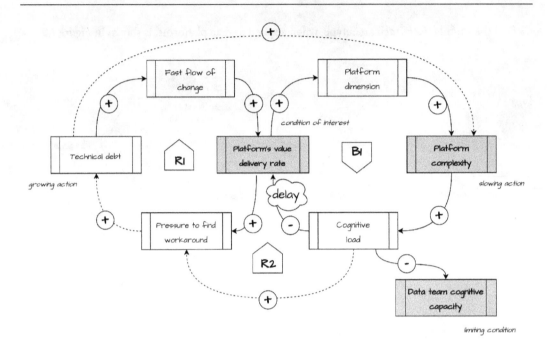

Figure 1.3 – Limits to growth

Externally, the delivery speed remains unchanged, but internally, the cognitive load on the development team grows rapidly until it saturates their capacity. At this point, efforts to maintain the past delivery speed become challenging. Shortcuts are no longer sufficient to compensate. The platform becomes a legacy that is difficult to evolve and costly to maintain. It's time to consider **replatforming** on a new technological stack. The solution to these challenges is modularization.

The **modularization** of the platform would reduce the cognitive load on the development team (technological architecture). Modularization also allows the cognitive load to be distributed across multiple independent teams (organizational architecture) if necessary. However, modularization is an activity that involves upfront costs with returns in terms of sustainability not being immediately evident. Modularization is also not a one-time activity but a process that must be carried out in tandem with the platform's evolution, slowing down its growth rate. Optimizing a system as a whole often involves balancing activities – in our case, modularization – that slow down growth to preserve sustainability.

> *"The optimal rate in a system is far less than the fastest possible growth. Optimal growth is the one that allows for the enhancement and resilience of the system, fostering a sustainable and harmonious balance over time."*
>
> *– Donella Meadows*

Without an understanding of the dynamics at play at the systemic level, obtaining the sponsorship and buy-in necessary to initiate such an activity at the organizational level becomes challenging. Why embark on a complex socio-technological transformation when we can solve the problems we've had so far using new, more innovative technologies (that is, shifting the burden)? Why deliberately slow down the delivery speed for potential medium-term sustainability benefits? We are aligned with current goals, and end-of-year incentives are tied to current performance, not to a promise of future value (that is, limits to growth).

The story of continuous stop-and-go over the last 30 years, a story dotted with new platforms built on the ashes of old ones unable to evolve, tells us that as difficult as it may be, it's time to change course and try new approaches to data management – approaches more focused on socio-technological architectures that address foundational problems condemning us to failure rather than seeking easy band-aid solutions in the latest technology trends. We can also consider approaches that allow us to break free from the chains of the failure cycles described in this section, in which we've been trapped like a hamster on its wheel for too many years. In this book, we will explore a possible way out centered around the idea of managing data as a product. But what does that exactly mean, and what is its relationship with what has been discussed so far? Let's explore this together.

Exploring why we need to manage data as a product

To escape the quagmire we find ourselves in, it is necessary to radically change the mental model we use to approach data management and, consequently, the organizational structures and associated operational practices. It's a systemic change – a paradigm shift in data management practice.

As we've seen in the previous sections, attempts to address these problems have been predominantly cosmetic, not radical. We've tried to modify the system tactically, reacting to surface-level problems as they arise.

In **system thinking**, a system can be changed from the outside by acting on parts of it where small changes can lead to significant and lasting changes over time; these parts are called leverage points. Donella Meadows, a renowned researcher in this field, has classified possible leverage points into 12 categories, ranking them by effectiveness. It's not necessary to delve into the details of each. Suffice it to know that the categories identified by Meadows can be summarized into four main macro-categories generally represented by a model known as the **iceberg model** (see *Figure 1.4*):

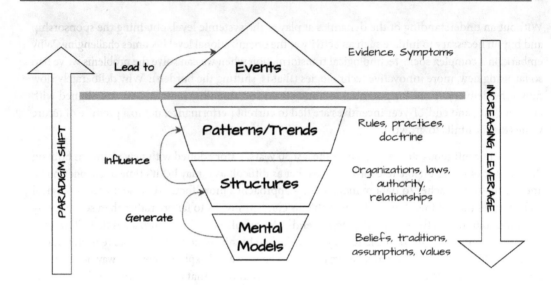

Figure 1.4 – The iceberg model

The visible events, and symptoms of deeper problems, on which we have predominantly focused until now, are the least effective leverage points. It's difficult to durably change the system by intervening on them. To make a qualitative leap in the system – that is, to make it capable of handling greater complexity – it is necessary to attack the levels of the iceberg that are below the waterline, outside the realm of immediately visible. The deeper you go, the more effective the leverage points become. A substantial change in the system's functioning coincides with a paradigm shift. It requires joint and coherent interventions on multiple leverage points, starting from the base of the iceberg (that is, mental models) and then moving up to higher levels (that is, organization and procedures). Let's see how.

Being data-centric

The *third industrial revolution* has led to a complete digitization of the core processes that govern our organizations. Applications, in addition to making processes more efficient, generate enormous volumes of data. This data has potentially immense value for organizations. It is a common belief that they represent a real asset, a strategic element to differentiate and compete in the market.

In 2016, Klaus Schwab, founder and executive chairman of the **World Economic Forum** (**WEF**), introduced the concept of the *fourth industrial revolution* to describe the period we are in now, a time of unprecedented changes and innovations reshaping the economic landscape. Data plays a fundamental role in this context. With many existing processes already digitized, the focus shifts to creating new ones, innovating, and evolving business models in directions unimaginable until a few years ago. And it is precisely here that data comes into play. Data is central to enabling these processes of innovative creation.

With the fourth industrial revolution, we are experiencing a transition from a model centered around applications, crucial for digitizing processes, to a model centered around data. The data produced by applications increasingly holds more value than the applications themselves. The lifespan of an application within an organization has significantly reduced. Applications today come and go under the pressure of constant technological innovation. Data, on the other hand, remains and is the true core asset for organizations.

While conceptually, the focus has shifted from applications to data, in practice, this transition has not fully taken place. Our mental models have not made this leap, and consequently, neither have the ways in which we manage data.

Applications still effectively drive the evolution of our IT architectures. Data outside of applications (data on the outside), despite being the real asset, is still treated as a second-class citizen. First, we choose applications, after which we understand what data they produce and how to integrate it with the information assets already available, not the other way around.

In summary, there is a disconnect between what we say when we talk about data and what we do in practice. It is essential to start from here, from the awareness of this fork, and adapt our mental model accordingly by placing data and its management at the center of the organization.

> *"Data is the center of the universe; applications are ephemeral."*
>
> *– The Data-Centric Manifesto*

Data is everybody's business

Moving up from the mental model to organizational structures, we can observe how data management in companies is heavily centralized in one or more dedicated teams. If we consider data management as a socio-technical system, data teams are parts of the system, while all other teams are not. The other teams, including those in IT managing infrastructure and applications, are part of the external context in which the system operates and interacts. This organizational model for distributing responsibilities related to data management conflicts with the mental model that places data at the center of the organization. It is primarily based on two mistaken assumptions. Let's see what they are.

The first assumption stems from a misinterpretation of data as an asset. If data is an asset, then it has value. If it has value, collecting more data adds more value, additively. However, data is a unique type of asset. Its value is not immediately fungible. It must be managed appropriately to effectively produce value. Data management is not comparable to refining crude oil. It is not a one-time transformation activity that turns raw data into valuable information. The value of data, even when properly transformed and enriched, tends to degrade over time. Therefore, data management is an ongoing effort. Given these premises, it becomes clear that thinking that accumulating data from sources is sufficient to generate value is incorrect. In reality, as we have seen, it is the opposite. As accumulated data grows, the system's complexity increases until a point is reached where management costs surpass the value produced, a point where the value of the collected data plummets (that is, limits to growth).

The second assumption is that managing data is primarily a technical problem. Perhaps it was true once. This was certainly the case with the advent of big data. Today, it is no longer the case. While technological complexity exists, it is a fraction of the total complexity related to data management. Today, the greater complexities lie in the organizational aspects linked to managing the data life cycle to maximize its value, reduce waste, and ensure sustainability over time. We should no longer think of the core capabilities of a mature data management practice purely in technical terms such as collect, process, and share. The core capabilities of modern data management should more appropriately be thought of in terms of identification, description, and standardization. The challenge is no longer collecting data but identifying which data is genuinely important and why. The challenge is no longer processing data but understanding how to describe it using models that can be functional to business needs. The challenge is no longer sharing data but standardizing the processes through which they are modeled and distributed to reduce waste, maximize interoperability, and encourage reuse.

All these challenges require active involvement in data management from the entire organization beyond the traditional roles of producer, processor, and consumer. On the one hand, application teams – traditionally data producers – must begin taking responsibility for the quality of the data they produce. On the other hand, the business – traditionally a consumer – must define which data is needed and what it means. These two activities, ensuring the quality of the source data and conceptually modeling the collected data, are currently the exclusive responsibility of data teams – teams that, on the one hand, do not have direct control over the source applications whose data quality they should ensure and, on the other hand, know little about the data that the business truly needs and its meaning. This situation only increases the complexity of management, the associated costs, and the speed at which the cognitive capacity of the data team is consumed. This latter variable is key to the sustainability of the entire system.

Figure 1.5 summarizes the key elements of the organizational changes described here:

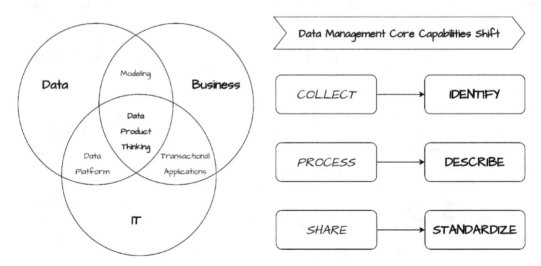

Figure 1.5 – Organizational changes

Now, let's move up from the organizational level to operational practices and see what it means to manage data as a product.

Data product thinking

Managing **data as a product** means applying product management practices to data management. Placing data at the center of the organization means no longer considering it just as a support component for products – that is, a byproduct of applications – but as the product itself.

Current monolithic platforms are more or less structured containers within which data is collected, processed, and redistributed according to project logic. When a new source system is released or when a business domain needs to produce new analyses, a project is initiated to add the necessary data to the platform. The platform is the product, and the data are features of the platform. Projects are how these features are implemented.

Managing a project is very different from managing a product. Projects have a budget associated with the solution they must develop and are evaluated based on their ability to release it on time and within the predetermined budget (that is, output). After the release, the team moves on to the next project based on portfolio priorities. Products, on the other hand, have a budget associated with a business problem or business capability and are evaluated based on the value produced in solving the problem or supporting the associated business capabilities (that is, outcome). Their development doesn't end with the first release but only when the product is retired. Throughout a product's life cycle, the budget dedicated to its development and maintenance can vary based on the relevance of the problem or business capability it's associated with. When it no longer has relevance, the budget is zeroed, and the product is retired. The evolution of a product is always overseen by the same team.

If product teams are not cross-functional from the beginning, they tend to become domain or problem experts. They become qualified and credible counterparts to external functions with which they must interact to advance the product's evolution. Project teams, on the other hand, tend to focus more on the precision of implementing requirements than on understanding the problem or domain for which the project is intended. While project teams are more focused on doing things right, product teams are more focused on doing the right things.

A product team can manage multiple products simultaneously. The only limit to the number of products a team can handle is its cognitive capacity. The number of product teams required depends on the organization's complexity and business needs. In simpler contexts, a single team may be able to manage all products. When multiple teams are necessary, they are generally associated with business domains. All products related to a specific domain are managed by the team associated with that domain. Over time, by becoming experts in the logic of the associated domain, the team can improve the quality of generated products and reduce the cognitive load for managing them. The budget associated with a product team is the sum of the budgets associated with the products it manages or the domain it refers to.

Data management is not a one-shot activity. Data has value and retains its value only if properly managed throughout its life cycle. Therefore, data is a product and should be managed as such. The data platform is built by composing data products that constitute its modular units. The teams developing the data platform are product teams associated with business domains, not project teams associated with technological components or logical layers of the platform.

Putting it all together

To prevent data platforms from collapsing under the weight of their complexity, it is necessary to reduce the cognitive load on the teams overseeing their development. The modularization of the architecture allows for a reduction in accidental complexity and, consequently, the cognitive load on the teams. Data products are the unit of modularization in the architecture we propose.

However, the practice of managing data as a product requires a broader paradigm shift to escape systemic dynamics that lead us to constantly repeat the same mistakes. Changing the paradigm means not only adopting new practices but also, and even before, changing the mental model and organizational structure. It is the mental model that supports the change in organizational structure, and the organizational structure, in turn, influences practices.

In this section, we discussed the need to shift from an application-centric view to a data-centric view at the mental model level. Then, we explored how, from a data-centric mental model, an organization can be derived, where responsibilities for data management are not concentrated but distributed. Only at the end have we seen how, from an organizational structure that promotes collaborative data management, the practice of treating data as a product can be derived.

To ensure the success of the paradigm shift, it is necessary to intervene synergistically at all levels, as depicted in *Figure 1.6*:

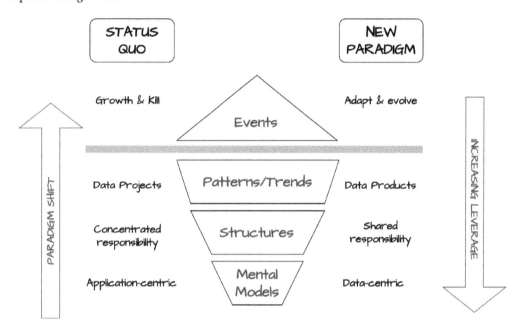

Figure 1.6 – Paradigm shift

Let's summarize the chapter now.

Summary

There is no doubt about the importance that data holds for organizations today to compete in the market. However, how we traditionally manage data has led to the construction of monolithic platforms unable to survive the complexity they generate.

Technological innovations have helped us solve many problems related to data management over time, but not the root causes of these problems. With each new generation of data platforms, the same issues have always resurfaced.

In this chapter, we saw how a paradigm shift in data management is necessary to make our platforms sustainable over time – a paradigm shift centered around the idea of treating data as a product and building modular platforms capable of governing the intrinsic complexity they generate without collapsing. It is a systemic transformation that touches on different levels of the organization. Throughout, we have shown the desired transformation required for each level, from mental models to operating practices.

In the rest of this book, we will explore through which leverage points we can intervene in the organization to promote the paradigm shift we have described here. The ultimate goal is to bring about a significant transformation across the organization, allowing us to build data platforms capable of producing sustainable value over time. In other words, we'll be transitioning from developing data platforms that follow a *growth and kill dynamic* to platforms that follow an *adapt and evolve dynamic*.

We'll start to embark on this journey by analyzing what exactly a data product is, and what its components and key characteristics are.

Further reading

For more information on the topics that were covered in this chapter, please take a look at the following resources:

- *An Architecture for a Business and Information System*, by B. Devlin and Paul T. (1988): `https://www.semanticscholar.org/paper/An-Architecture-for-a-Business-and-Information-Devlin-Murphy/c22ce1eeafb01f0682e194a2a22349aa141b78f6`

- *Building the Data Warehouse*, by W. H. Inmon (1992): `https://www.amazon.com/Building-Data-Warehouse-W-Inmon/dp/0764599445`

- *The Data Warehouse Toolkit: The Definitive Guide to Dimensional Modeling*, by R. Kimball and Margy Ross (1996): `https://www.amazon.com/Data-Warehouse-Toolkit-Definitive-Dimensional/dp/1118530802/`

- *Data Lake Architecture: Designing the Data Lake and Avoiding the Garbage Dump*, by W. H. Inmon (2016): `https://www.amazon.com/Data-Lake-Architecture-Designing-Avoiding/dp/B01HN4JOPC/`

- *The Modern Data Stack: Past, Present, and Future*, by Tristan Handy (2020): `https://www.getdbt.com/blog/future-of-the-modern-data-stack`

- *Design Rules: The Power of Modularity*, by C. Y. Baldwin and K. B. Clark (2000): `https://www.amazon.com/Design-Rules-Vol-Power-Modularity/dp/0262024667`

- *Thinking in Systems*, by D. H. Meadows (2008): `https://www.amazon.com/Thinking-Systems-Donella-H-Meadows/dp/1603580557`

- *The Fifth Discipline: The Art & Practice of The Learning Organization*, by Peter M. Senge (2006): `https://www.amazon.com/Fifth-Discipline-Practice-Learning-Organization/dp/0385517254`

- *The Fourth Industrial Revolution*, by K. Schwab (2017): `https://www.amazon.com/Fourth-Industrial-Revolution-Klaus-Schwab/dp/1524758868`

- *The Data-Centric Manifesto*: `http://datacentricmanifesto.org/`

2

Data Products

In the previous chapter, we explored the importance of managing data as a product rather than simply as a by-product. Managing data as a product involves applying the principles of product management to data management. These principles are better suited for handling a strategic asset such as data compared to the principles of project management.

In this chapter, we will delve into what exactly a data product is by first introducing the concepts of data product, pure data product, and analytical application. We will then define what a pure data product is and proceed to analyze its structure and key characteristics. Finally, we will examine some commonly adopted methods for classifying pure data products.

This chapter will cover the following main topics:

- Defining a data product
- Exploring key characteristics of pure data product
- Dissecting the anatomy of a pure data product
- Classifying pure data product

Defining a data product

Managing data as a product is a common principle in many modern approaches to data management. In recent years, much has been said about the implications of this practice and what a data product is or should be.

While there is a broad consensus on many characteristics that a data product should have, there is not as much agreement on a shared definition of what a data product is. DJ Patil, the former chief data scientist of the United States, defines a data product as follows:

A data product is a product that facilitates an end goal through the use of data.

The definition is clear, but it's quite broad. Despite its ready acceptance, the absence of specificity detracts from its practical utility in a hands-on context.

In this section, we will attempt to expand DJ Patil's definition to arrive at a more actionable definition of a data product, which we will then use as a reference throughout the rest of the book.

Pure data products versus analytical applications

According to the definition by DJ Patil, every digital application developed following the principles of product management (i.e., digital product) could be considered a data product. Every digital application, to achieve its goals, makes use of data. The technologies used to develop digital applications are called information technologies for this very reason. This definition doesn't dissatisfy anyone, but it's too general to have practical value. It needs to be expanded and made more specific. Let's see how.

In general, we can divide digital applications into two major categories: **transactional applications** and **analytical applications**. For the former, functionalities drive data management, while for the latter, data guides the development of functionalities. In transactional applications, data is essentially a supporting component of the product, a by-product. In analytical applications, however, data is a core component of the product. For this reason, it makes sense to not consider transactional applications as data products, extending the definition provided by DJ Patil in the following way:

"A data product is a product in which the use of data not only supports *but also drives the development of the functionalities* necessary to achieve its goals."

Data is a core component of analytical applications. By adhering to product management principles during development, analytical applications qualify, according to the previous definition, as data products while transactional applications do not. Let's now introduce the concept of a **pure data product**, which will be the focus of the rest of the book. A pure data product is a new type of digital product, setting aside transactional and analytical applications, whose goal is to expose high-quality data in its pure form to support the development of other applications. For a pure data product, data is not just a core component for achieving its objectives; it is the objective itself. The data is the product.

While an analytical application delivers functionalities that use data, a pure data product directly delivers the data. When we think of an analytical application, we generally think of reports and dashboards. These types of applications do not directly expose the data but rather a visual representation of the data. An analytical application, such as a report or a dashboard, can, therefore, be considered a data product but not a pure data product.

Overt data products versus covert data products

Mike Loukides, vice president of content strategy for O'Reilly Media, introduced the concepts of **overt data products** and **covert data products**. In an overt data product, the data is clearly visible as part of the deliverable. On the contrary, the data is hidden and not visible to the end user in a covert data product. Pure data products are, by definition, overt data products. Analytical applications, however, fall on a continuous spectrum based on how much the data is hidden from the end user by the exposed functionalities. A report is an example of an analytical application where the data is prominent, making it an extremely overt data product. On the other extreme, a recommendation system is an analytical application in which the data determining its functionality is completely invisible to the user, making it an extremely covert data product.

The classification of data products described so far is shown in the following figure (*Figure 2.1*).

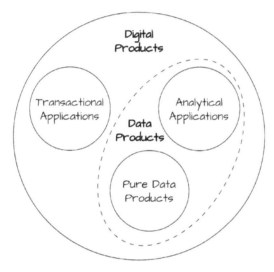

Figure 2.1 – Classification of digital products

We have chosen to separate pure data products from analytical applications and not consider the former as a subset of the latter because the goals of the two types of products are not only different but almost entirely orthogonal. Pure data products aim to expose data to support multiple use cases, both of analytical and transactional nature. Their value is greater when the data is of high quality, easily accessible, and reusable over time. Analytical applications, on the other hand, are designed around a specific use case. Their value is greater as they better support the use case for which they are designed. Generally, the more hidden the data, the easier and more frictionless the analytical application supports the user. While pure data products focus on solving data problems, analytical applications focus on solving business problems. The former are completely overt data products because their users need data. The latter naturally tend to be covert data products because their users don't need the data but rather actionable results.

Why do we need pure data products?

Data management practitioners are often obsessed with the idea of having to prove the value they bring to the business. There isn't a data conference, talk, or podcast where the topic of the value generated by data management initiatives and how to prove it isn't extensively discussed.

Of course, all business functions must deliver value to the organization. However, for data management, this seems to be an existential problem, as proving the generated value is essential to justify the existence of the function itself. No one questions whether the marketing, sales, or HR functions make sense based on the value they produce for the company. The fact that these functions are necessary is commonly accepted. When we talk about value, we are referring to the performance of individual teams, understood as their ability to contribute to the goals assigned to them based on the company's strategy.

The fact that data management teams feel the need to prove not only their performance but also the need to exist as a business function indicates a problem. The causes are mainly those described in the previous chapter. The periodic rebuilding of data platforms on new technologies, motivated by the collapse of the previous ones, has led to significant investments, often only partially justified by returns. The business is rightly hesitant to approve further investments in technology in the hope of solving problems that seem chronic in the data function.

In this context, to approve new initiatives, it is increasingly necessary to provide clear evidence of the value delivered in relation to the required investments. However, we should not confuse the value produced for the individual user or business function with the value produced for the overall organization. The value produced for the user or business function is linked to local needs, often of a tactical nature. The measurement of the produced value is mostly based on **leading indicators**, which are immediately sensitive to the initiatives implemented (e.g., the number of data assets managed, number of reports implemented, number of potential users, etc.). On the other hand, the value produced for the business is linked to global needs, often of a strategic nature. The measurement of the produced value is mostly based on **lagging indicators**, which are systemic and tend to vary over longer timeframes (e.g., maintenance cost, average time and costs to produce new analyses, active users, users' trust in the data accuracy, etc.).

Focusing on the needs of individual users or groups of business users allows for quicker evidence of the delivered value. However, completely ignoring the value delivered for the organization as a whole puts the system at risk. Accounting for the generation of value at the local level tends to overlook the lagging indicators typical of global value generation. These are generally indicators whose function is to ensure strategic coordination and sustainability over time. Therefore, a holistic approach to defining the produced value is necessary to balance local needs with global needs, as shown in the following figure (*Figure 2.2*). Only in this way, the organization can sustainably evolve toward its purpose.

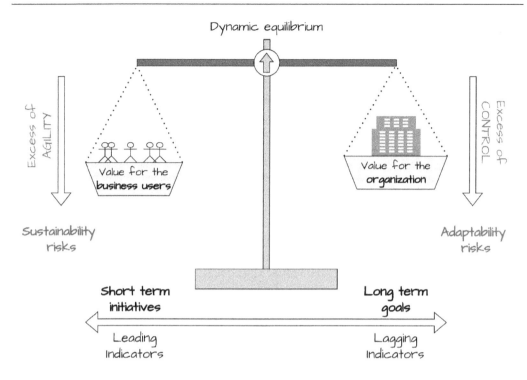

Figure 2.2 – Short-term and long-term value

In the field of data management, this means advancing initiatives that address the needs of business users while simultaneously not neglecting the overall needs of the organization. No shortcuts, in other words.

A set of **metrics of success** (**MoS**) for holistically evaluating each initiative is necessary. A set of MoS, such as net promoter score of technical and business users, management costs, change failure rate, MTTD, and MTTR, is adept at harmonizing lagging and leading indicators, ensuring that local initiatives do not erode the long-term sustainability of the entire system.

This holistic approach has been lacking over time and has often led to the collapse of data platforms. As explained by the **limit to growth** archetype, described in *Chapter 1, From Data as a Byproduct to Data as a Product*, the pressure to quickly meet the needs of business users has led to neglecting the sustainability of the platform over time, introducing **technological debt** and thus increasing management complexity. The problem is not that we measure value based on lead time. The problem is that only this leading indicator is used without counterbalancing it with lagging indicators such as management costs. The negative externalities resulting from technological debt do not appear in the set of MoS. Consequently, no guardrails are limiting technical debt production, but it is the technological debt that ultimately undermines the sustainability of the platform as a whole.

Returning to data products, we have seen how analytical applications focus on business problems or use cases of interest to business users. Pure data products, on the other hand, focus on relevant data problems. While analytical applications directly generate value for business users measurable with leading indicators, this is not possible for pure data products. Many might be tempted to doubt the need for pure data products because it's more challenging to demonstrate immediate value to the business. No business user directly accesses a sales dataset. Business users access a report that shows sales trends. It's the report that produces value for them, not the dataset.

So, why define pure data products as entities separate from the analytical applications that use them? Isn't it sufficient to consider them as a component of the latter and focus only on developing analytical applications whenever there are business use cases? The answer, based on what has been said so far, is negative. Let's see why in more detail.

Developing analytical applications solely based on the use cases that arise corresponds to an approach to production called **engineer to order** (**ETO**). In this approach, each product is designed, engineered, and built according to the specific requirements of a customer. The focus is almost exclusively on maximizing the value produced for the specific customer, while little or no investment is made in optimizing the production process. This approach is common in sectors where each product is a unique piece, and there is little room to reuse its components to build other products. ETO is applied for example in the production of industrial machinery (e.g., packaging machines), custom architectural structures (e.g., bridges and overpasses), specialized vehicles (e.g., emergency response vehicles), and high-end custom electronics (e.g., specialized medical devices).

This is not the case with data management. Each analytical application has the task of collecting data from one or more domains, consolidating them, and enriching them in a way that is functional to the use case it must support. However, the data mostly come from the transactional systems that govern the company's core processes. The data used by an analytical application is therefore certainly also used by other analytical applications. In an ETO approach, however, the integration of this data must be managed independently by each analytical application. This means satisfying users at a local level but losing the opportunity at a global level to pool integration costs, reduce maintenance costs, and minimize risks of misalignment in data management.

Applying the ETO approach to data management means keeping data in silos not only at the level of transactional applications but also at the level of analytical applications, as shown in the following figure (*Figure 2.3*).

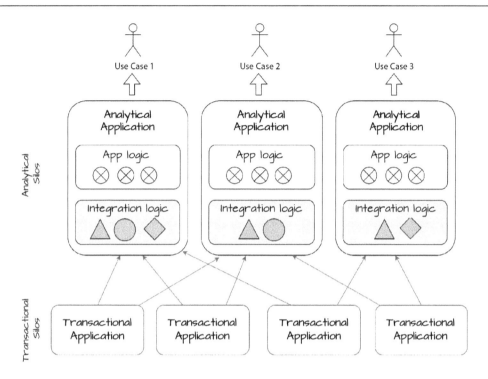

Figure 2.3 – ETO

In data management, it is preferable to adopt an **assemble-to-order** (**ATO**) production approach, where products are built on request by assembling pre-existing components. ETO is applied, for example, in the production of personal computers, automobiles, and consumer electronic devices.

Applying ETO to data management means outsourcing the data integration work from analytical applications. Each analytical application can collect the necessary data from components dedicated to this purpose, the pure data products. Pure data products, as depicted in the following figure (*Figure 2.4*), are positioned between transactional applications and analytical applications to rationalize data integration costs and facilitate the development of new use cases.

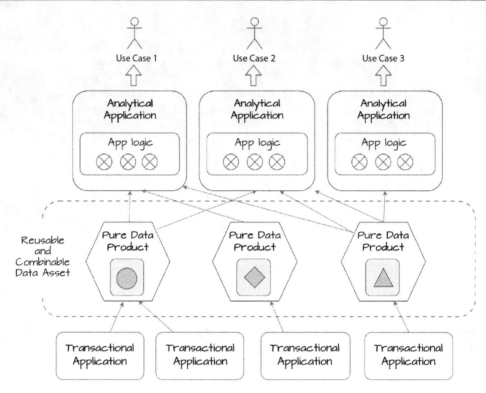

Figure 2.4 – ATO

In summary, while analytical applications are necessary to meet user needs, pure data products are essential to address the scalability and sustainability requirements of the organization on a global scale.

We've seen that the development of analytical applications tailored to specific use cases, in the absence of a foundation of pure data products, can lead to inefficiencies that, in the long run, may create systemic issues. The same holds true for the development of pure data products without real use cases, following a **make-to-stock** (**MTS**) production approach. MTS is applied, for example, in the production of standardized clothing, canned goods, household products, and medicines.

Creating and maintaining pure data products over time has a cost, but their value is realized only when they are used. Developing pure data products without real use cases following the MTS approach to production is a waste of resources. The correct approach is to prioritize the development of pure data products based on the needs of the use cases that will be implemented. It is the use case that determines which pure data products to develop or extend. However, the use case does not dictate how the pure data product will be developed or extended. Pure data products seize the opportunity provided by the use case to integrate one or more data assets and make them easily usable for a wide range of present and future use cases.

Pure data product definition

Now that we've delved into the concept of a pure data product, and justified its importance, it's time to articulate a more precise definition. Here it is:

"A *pure data product* is a modular unit within the data architecture, tailored to the cognitive capacity of the responsible team and developed following product management principles to make one or more data assets accurate, relevant, combinable, and readily usable for future value creation."

Let's unpack the definition into its key elements:

- A pure data product is a unit of modularization within the data architecture (**architectural quantum**)
- The complexity of a pure data product should not surpass the cognitive capacity of the responsible team (**cognitive fit**)
- A pure data product is developed following the principles of product management (**data as a product**)
- The objective of a pure data product is to make one or more data assets readily usable in multiple use cases (**liquid data assets**)

Architectural quantum: A pure data product is a unit of modularization within the data architecture. It is the smallest module that can be managed independently during its life cycle. It may have internal components, but they share its life cycle.

A pure data product is a module because its internal components exhibit a high level of cohesion and are loosely coupled with external components and modules. It is a module also because interactions with the external world occur through interfaces (abstraction) that mask the functioning of its internal components (information hiding). A pure data product is not just a module anyway. It is a unit of modularization because it cannot be decomposed into sub-modules that are independent of a life cycle management perspective. Sub-modules lack the level of autonomy, encapsulation, and information hiding necessary to be considered pure data products themselves:

Cognitive fit: The dimensions of a pure data product must not exceed the cognitive capabilities of the team responsible for it. One of the main reasons for shifting from monolithic data architectures to modular architectures is precisely an attempt to reduce the cognitive load on development teams to overcome the systemic issues described in *Chapter 1, From Data as a Byproduct to Data as a Product* through the archetype of limits to growth. It is therefore important to keep the complexity level of each pure data product in check to avoid reproducing the problems of monolithic architectures within pure data products over time. It is crucial to avoid creating monolithic pure data products whose complexity makes them challenging to manage and evolve. When a pure data product grows beyond the cognitive capabilities of the responsible team, it is advisable to initiate refactoring to break it down into smaller data products.

Data as a product: A pure data product is developed following the principles of product management. It is not a project deliverable aimed at implementing a set of predefined requirements. Instead, a pure data product is associated with a business-relevant data problem. It has a dedicated team that oversees its development and evolution over time, optimizing the associated budget to create a product that represents the best possible solution for the target problem.

Liquid data asset: The goal of a pure data product is to make a data asset ready to be used in an unbounded set of use cases. Just like any other business asset, data comes with a cost tied to its management and a value associated with the utility it provides for the organization. However, the parallels stop there.

Professors Daniel Moody and Peter Walsh from the University of Melbourne have identified the following seven laws that govern the value of data assets, making them different from traditional assets:

- Information is (infinitely) shareable
- The value of information increases with use
- Information is perishable
- The value of information increases with accuracy
- The value of information increases when combined with other information
- More information is not necessarily better
- Information is not depletable

Therefore, data should not be treated like any other asset. Data is unique; it can be freely reused and recombined countless times without losing its value. On the contrary, its value grows when reuse is encouraged, reducing the copies that only increase management costs. Data assets, as observed by the **MIT Center for Information Systems Research (MIT CISR)**, are born to be liquid. The higher their ease of reuse and recombination, the more liquid they become. To fully capitalize on the potential value of data, it's essential not to confine them to supporting specific use cases (analytical applications) alone. Instead, they should be managed to make them liquid or, in other words, to make them accurate, relevant, combinable, and readily usable for future value creation. The capability of making data assets liquid is one of the most important characteristics of a good pure data product.

To ensure that data assets, managed by a pure data product, are truly ready for the implementation of new use cases, reusability and combinability alone are not enough. Accuracy and relevance are also needed. The value of data assets grows with their accuracy. Below a minimum threshold of accuracy, a data asset has no value. Data assets also tend to devalue over time if not properly maintained. It is therefore necessary to work at the product level to preserve that relevance over time. Since the development and evolution of a product have a cost, it is also important to identify from the beginning the data assets to be managed based on their relevance to the business. This process of identification and selection is important not only to optimize management costs but more generally to increase the return on investment made in pure data products. Contrary to other assets, the value of data does not grow in relation to quantity. When managed data becomes too much, its value tends to decrease due to the phenomenon of **information overload**. In other words, when the quantity of available data saturates the cognitive capabilities of potential consumers, their ability to understand and use the data is reduced. The value derived from adding new data becomes marginal.

The rise of data-driven applications

The evolution of technologies is transforming the data value chain from a linear to a circular value generation process where the roles of data producers and data consumers are increasingly blurred. In this context, a new type of application is emerging—data-driven applications. Let's delve deeper into what these applications entail and examine the dynamics driving this transformation in the data value chain.

Traditionally, transactional applications generate data that analytical applications then use to extract various insights. The data value chain is linear and unidirectional. It starts from transactional applications that generate raw data and ends after several steps of consolidation, cleaning, and enrichment in analytical applications that produce actionable insights. Data transformations predominantly occur in batch mode, and the produced insights are primarily descriptive and diagnostic. In this scenario, pure data products are positioned in the middle, between transactional and analytical applications. The former generates data that pure data products consolidate and make available for consumption by the latter. The introduction of pure data products modifies the socio-technological architecture of the system but does not alter the structure of the data value chain. The data continues to flow unidirectionally from transactional systems to analytical systems, as shown in the following figure (*Figure 2.6*).

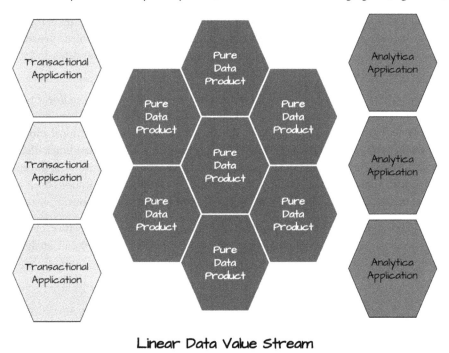

Figure 2.5 – Linear data value stream with pure data products in the middle

This unidirectionality is now being questioned by technology innovations and in particular by the following:

- The evolution of data integration technologies
- The increasing use of AI in the development of analytical applications

Evolution of data integration technologies

The evolution of data integration technologies increasingly allows for the reduction of latencies in which data is extracted from transactional systems, processed, and made available to analytical systems. In addition to traditional batch integration methods, streaming methods based on events are becoming more prominent. In some cases, the latter has even become the main integration method, with batch integration being a particular case. In this scenario, pure data products expose enriched and curated data in near real time, with minimal latencies compared to when the raw data on which the enrichment is based is inserted into source transactional systems. This makes pure data products usable not only by analytical applications but also by transactional applications. Transactional applications are thus both data producers used by pure data products and potential consumers.

Increasing use of AI in analytical applications

The increasing use of AI in analytical applications allows the generation of not only traditional descriptive and diagnostic insights but also predictive and prescriptive insights. The latter makes the time from insight to action faster, not requiring or significantly reducing direct human intervention. Insights can be quickly retro-propagated to transactional systems to optimize their processes. This forms a value cycle, where data flows from transactional systems to analytical systems and vice versa in the form of actionable insights, known as **continuous intelligence**.

Thanks to these technological innovations, the data value stream is shifting from linear to circular. The categorization of transactional applications as data producers and analytical applications as data consumers becomes blurred. In many cases, even the distinction between transactional and analytical applications tends to blur. More and more, analytical applications are seen providing a high level of user interaction, supporting transactional workloads, and vice versa, transactional applications are more equipped with analytical functionalities. We call this new family of applications born from the convergence of transactional and analytical applications **data-driven applications**.

In this scenario, pure data products are positioned at the center of the architecture, ensuring the availability of accurate, relevant, composable, and easy-to-use data for transactional, analytical, and data-driven applications (*Figure 2.6*).

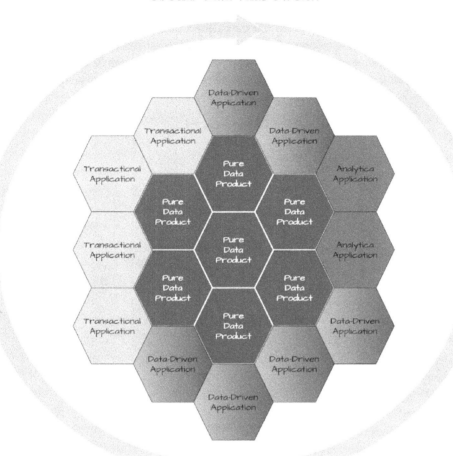

Figure 2.6 – Circular data value stream with pure data products in the center

Now we know what a pure data product is and how it fits into IT architecture and value generation processes. But what are the key features for measuring the quality of a pure data product? Let's explore them together in the next section.

Exploring key characteristics of pure data product

In the previous section, we explored how a pure data product aims to make one or more data assets relevant, accurate, reusable, and composable. In this section, we will delve more deeply into these characteristics to better understand how to assess the quality of a pure data product and the design strategies to follow during its development to ensure that quality.

Popular ilities

The non-functional characteristics that determine the quality of an application are generally referred to as **ilities** due to the suffix that many of the words used to define them share (e.g., availability, scalability, reliability, etc.). The key *ilities* we have identified for a pure data product stem directly from its raison d'être, namely, to maintain (relevance and accuracy) and enhance (reusability and composability) the value of one or more data assets (*Figure 2.7*).

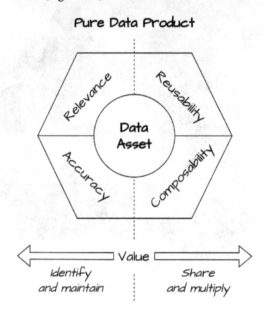

Figure 2.7 – Pure data product ilities

A lot has been discussed recently about the characteristics to use for assessing the quality of a data product. The scientific community originally proposed and promoted four key characteristics identified by the acronym **FAIR**, which stands for **findability, accessibility, interoperability, and reusability**, intending to maximize the utility of data, promote collaboration, and foster the advancement of knowledge and innovation. The FAIR characteristics focus exclusively on aspects that enable the growth of the value of data through sharing. However, they do not cover the part related to identifying the most valuable data and the ability to sustain this value over time.

Later, several other proposals emerged to broaden the FAIR characteristics, tailoring them to the specificities of data products and extending beyond aspects solely tied to data sharing. Undoubtedly, Zhamak Dehghani's proposition stands out as the most popular. Dehghani identifies the following eight non-negotiable basic characteristics that a data product must have to be considered useful: discoverability, addressability, understandability, trustworthiness, native accessibility, interoperability, valuable on its own, and security.

Table 1.1 shows how the characteristics identified in the definition of a pure data product proposed here map to those of FAIR and those defined by Dehghani.

Pure data product	FAIR	Data Mesh	Why?
Relevance	-	Valuable on its own	For the following reasons: Data is perishable More data is not necessarily better
Accuracy	-	Trustworthy	For the following reasons: The value of data increases with accuracy
Reusability	Findability, accessibility, reusability	Discoverability, addressability, understandability, native accessibility, secure	For the following reasons: Data is (infinitely) shareable The value of data increases with use Data is not Depletable
Composability	Interoperability	Interoperability	For the following reasons: The value of data increases when combined with other data

Table 1.1 – Pure data product key characteristics

The last column of the table shows how each characteristic can be associated with one of the seven laws that contribute to determining the potential value of a data asset according to Moody and Walsh.

Now, let's delve into the details of each characteristic that determines the quality and potential value of a pure data product.

Relevance

The potential value of a data asset is closely tied to its relevance for its users. The relevance of a data asset is extremely subjective and can vary over time. A certain piece of data may have immeasurable value for some users and be insignificant for others. Similarly, data that is relevant today may lose its relevance in the future. We have already observed that attempting to manage all data generated by an organization through pure data products, besides being economically unsustainable, does not yield value beyond a certain volume of data managed due to the phenomenon of information overload (*Figure 2.8*).

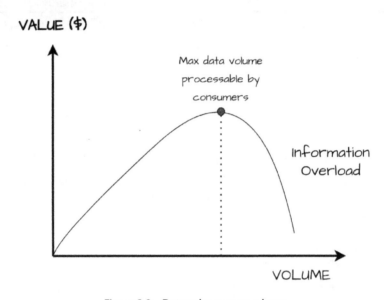

Figure 2.8 – Data value versus volume

It is therefore important to carefully identify the most relevant data for the organization on which to build pure data products. It is also crucial to continuously monitor their relevance over time to decide whether and how to modify the investments within the product portfolio. When the business relevance of a product falls below a certain threshold (i.e., the product is not used anymore to implement new use cases), the product enters maintenance mode. When it becomes irrelevant (i.e., the product does not support anymore any valuable use case), it should be decommissioned to reduce the complexity of the data landscape and the informational burden on potential consumers. This activity of identifying relevance is a task that must necessarily be carried out by business people per the corporate strategy.

Moreover, unlike other assets, data is perishable, meaning it tends to lose value over time. Some data degrades more rapidly than others. In general, the fresher the data, the more relevant it is; conversely, as it ages, it loses relevance and, consequently, value (*Figure 2.9*).

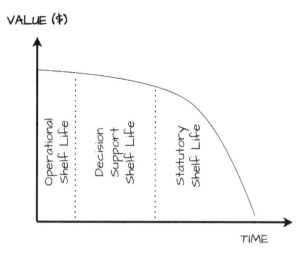

Figure 2.9 – Data value versus time

In this case, it is the responsibility of the product team to ensure a continuous update of data, preferably with low latencies, to always have fresh and relevant information. The product team always must ensure that stored historical data is removed when it falls out of the relevance window, making its management only a cost without generating value.

Finally, it is important to emphasize that the relevance of data should be understood as its significance for the organization that owns that data. In essence, it is relevant for the business, not for the IT teams managing the data. A pure data product can be combined with other pure data products to enable the implementation of a business use case. However, it should not always be necessary to combine it with other data products for it to be relevant and usable for implementing a business use case. If this happens, the pure data product is not relevant by itself for the business. It would probably be better to incorporate it as a component of another pure data product. It may seem obvious, but it is a common mistake when coming from a layered data architecture (e.g., medallion architecture) to define pure data products as sub-modules of existing integration pipelines, aligned with consolidation layers and their respective technologies rather than potential business areas of interest. It erroneously prioritizes functional relevance to IT operations rather than functional relevance to business activities.

 A data asset exposed by a pure data product is relevant if it is the following:

* **Business-aligned**: The data addresses high-priority business needs
* **Timely updated**: The data is continuously updated, preferably with low latencies
* **Non-redundant**: Collected data that is no longer needed is periodically eliminated

Accuracy

In addition to being relevant, data must be accurate. The level of accuracy required depends heavily on the type of data and the use cases one aims to support. In general, the more accurate the data, the greater the number of potentially supported use cases. However, depending on the type of data, beyond a certain level of accuracy, as shown in the following figure (*Figure 2.10*), the value obtained from further increasing accuracy becomes marginal. In other words, it does not enable additional significant use cases.

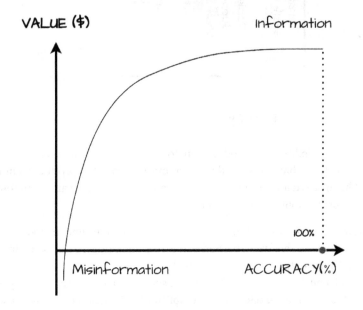

Figure 2.10 – Data value versus accuracy

Below a minimum threshold of accuracy (i.e., when the quality of exposed data is not sufficient to support the use cases at hand), data management becomes a cost that doesn't generate value but creates risks. It is the responsibility of business people to define the necessary accuracy level for a specific data asset to support the use cases of interest. On the other hand, it is the responsibility of the product team to ensure the desired level of accuracy.

A data asset exposed by a pure data product is accurate if it is the following:

- **Complete**: All relevant data is present
- **Consistent**: The data is uniform and does not contradict itself
- **Valid**: The data adheres to predefined rules and constraints
- **Trustworthy**: The data is perceived as accurate by potential consumers
- **Precise**: The data is collected at the right level of detail

Reusability

Relevance and accuracy are essential characteristics for identifying valuable data assets and maintaining their value over time. Reusability and composability, on the other hand, are indispensable characteristics for sharing and multiplying the value of a data asset. Unlike other assets, data can be used simultaneously by multiple consumers without losing value. As illustrated in the following figure (*Figure 2.11*), they gain more value as they are used more frequently.

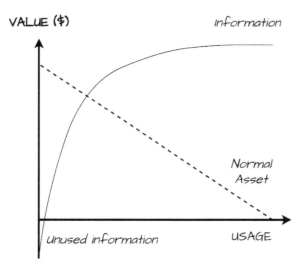

Figure 2.11 – Data value versus usage

The potential value of a data asset is at its peak when everyone within an organization is aware of its existence, understands its significance, can easily access it based on their role, and knows how to use it effectively. It is the responsibility of the product team to ensure the reusability of data assets exposed by the pure data product.

A data asset exposed by a pure data product is reusable if it is the following:

- **Discoverable**: The data can be easily found
- **Addressable**: The pure data product has a globally defined unique and persistent identifier that allows potential consumers to locate and access its data and metadata
- **Understandable**: The syntax and semantics of the data are clear and comprehensible to potential consumers
- **Accessible**: Data can be programmatically accessed through stable and documented interfaces
- **Reliable**: The data product exposes data according to the declared **service-level agreements (SLAs)**
- **Secure**: Data is available for reuse in compliance with security policies and regulatory constraints applicable to the organization (e.g., GDPR)

Composability

Data assets gain value not only when used individually but also when combined with other data assets to produce new enriched data or to support new business use cases (*Figure 2.12*).

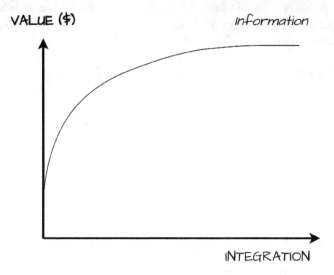

Figure 2.12 – Data value versus integration

To combine data assets exposed by different pure data products, it is necessary to globally define shared governance rules to ensure interoperability. This involves establishing standards, protocols, and policies that enable consumers to easily and correctly link different data assets at both the syntactic and semantic levels. Without these global governance rules, the independence of pure data products quickly turns into anarchy, transforming the platform into a collection of data silos and placing the entire complexity of integrating different data assets on data consumers.

A data asset exposed by a pure data product is composable if it is the following:

- **Technically interoperable**: The pure data product exposes data in compliance with standard communication protocols defined at the global governance level

- **Syntactically interoperable**: The pure data product exposes data and metadata in compliance with formats defined at the global governance level

- **Semantically interoperable**: The pure data product links the exposed data with terms from the enterprise business glossary or concepts from the enterprise ontology

- **Reasonably stable**: The public interface of the pure data product (data contracts) remains relatively stable over time

So far, we've seen what a pure data product is and what its characteristics are. We've essentially analyzed a pure data product only from the outside. In the next section, we'll analyze its internal structure.

Dissecting the anatomy of a pure data product

In this section, we will delve into the anatomy of a pure data product by breaking it down into its main components and then analyzing each of them individually.

Anatomy overview

A pure data product, as a unit of modularization of the data architecture, must contain all the components necessary for its proper functioning. It should include all the elements that enable it to be deployed initially and then managed throughout its life cycle independently (architectural quantum).

Data assets produced for consumption are not pure data products in themselves. They are a part of it. Alongside managed data, the other key components constituting a pure data product include, as illustrated in the following figure (*Figure 2.13*), the metadata associated with the product, the infrastructural resources used, the applications that transform the data and expose services to external consumers, and the interfaces through which data and services exposed by the product can be consumed.

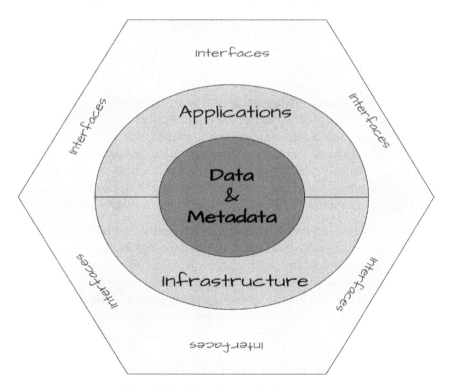

Figure 2.13 – Pure data product's anatomy

Let's analyze each component in detail.

Data

A pure data product acquires data from external systems or other pure data products, transforms them, and makes the resulting data available for consumption. Only the transformed data is accessible externally (information hiding). Consumption never occurs through direct access to the data but is always mediated by interfaces (abstraction).

A pure data product should expose only enriched data, minimizing the exposure of data acquired from other pure data products. Deploying data already exposed by other pure data products is equivalent to creating a copy of this data, which, as seen before, increases overall management costs without adding value to the data itself. Furthermore, exposing the same data to different products creates ownership issues and potential risks of inconsistencies.

For example, consider a pure data product that acquires customer data from another product to calculate a scoring index (e.g., churn probability). The data exposed by this data product should be limited to the customer's identifier and the calculated score. It should not expose, through mere copying, other customer data whose ownership belongs to another product. The merging of the scoring index with other customer data of interest should be done directly by the consumer, preferably with the support of a data virtualization tool that facilitates the execution of federated queries among data exposed by different products.

Metadata

The **metadata** associated with a pure data product goes beyond describing the exposed data; they provide a comprehensive description of the product as a whole. Metadata describing a pure data product (e.g., identifier, name, domain, owner, purpose, etc.), its interfaces (APIs, service levels, usage terms, etc.), and the exposed data (schema, constraints, semantics, etc.) are public. In other words, they are visible to all potential consumers. On the other hand, metadata describing its internal components (applications and infrastructure) are private. In other words, they are visible only to the product team and potentially to the tools used to automate the product life cycle management.

All metadata associated with a pure data product should be contained in a **data product descriptor document** defined according to a specific, formal, and machine-readable standard specified at the organizational level. The part of the descriptor containing public metadata, accessible to all potential data consumers, is generally referred to as the **data contract**. The data product descriptor serves as the *lingua franca* for exchanging information about pure data products among the products themselves, consumers, and supporting tools. It is a key element in promoting interoperability between different pure data products. We will delve into the structure and methods of defining a data product descriptor in *Chapter 5, Designing and Implementing Data Products*.

It's important to note that the responsibility for managing metadata related to a pure data product falls entirely on the product team. Compared to traditional data governance models, the data product-centered approach shifts left metadata responsibilities, transferring ownership from the central data stewards' team to product teams. Metadata is an integral part of the pure data product and shares its life cycle. In data and metadata management, there is neither spatial separation (ownership by different teams) nor temporal separation (data released first, metadata later). A pure data product cannot be released without a descriptor document containing all necessary metadata. Defining all required metadata in the descriptor document is part of the **definition of done** for a pure data product. The enforcement of this policy on metadata should be included in the products' automated procedure of **continuous integration and continuous delivery** (CICD).

Application and infrastructure

A pure data product is also composed of applications and infrastructural resources. The applications implement the necessary business logic to transform the data (data pipelines) and, more generally, to provide all the services exposed by a pure data product through its interfaces. The infrastructural components, on the other hand, provide the resources needed for applications to store and process data. Applications and infrastructural resources are not visible to consumers, who only see the public interfaces. If these interfaces remain unchanged, applications and infrastructural resources can be freely modified over time without impacting consumers.

Pure data products can use external and shared infrastructure services and resources. It's anyway crucial that these external services and resources allow segregated usage and do not contain business logic. It is important that the use of external components, both at the application and infrastructure levels, enables rationalization of technological costs without creating dependencies between products that undermine their autonomy. It is perfectly normal and common, for example, to have multiple products sharing the same storage infrastructure (e.g., analytical database). The crucial point is that they do not directly share data at the storage level (e.g., tables). Sharing must always occur through interfaces. If the storage infrastructure is shared, each product must autonomously manage the resources it creates within this infrastructure, and the infrastructure must ensure that other products cannot access them directly (multi-tenancy).

Interfaces

A pure data product exposes data and functionalities through interfaces that allow external consumers to abstract from the internal implementation logic. The goal of a pure data product is to expose one or more data assets to potential consumers, promoting reuse and composability. The interfaces used to acquire and expose data play a key role but are not the only ones that a pure data product can make available to its consumers. Typically, a pure data product also provides functional interfaces to understand, monitor, and control its behavior. The interfaces exposed by a pure data product are usually referred to as **ports** and grouped by functional purpose. The main types of ports that a data product can have, as shown in the following figure (*Figure 2.14*), are **input ports**, **output ports**, **discoverability ports**, **control ports**, and **observability ports**.

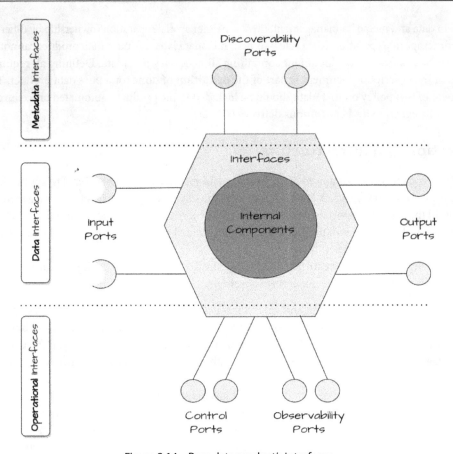

Figure 2.14 – Pure data product's interfaces

Let's go through these types of ports that a data product can have in detail:

- **Input ports**: These specify how and from where the pure data product acquires data, which it then enriches and exposes to its consumers. Sources can be output ports of other products or external transactional systems. The acquisition can occur in push or pop mode. In the former, the input port describes the API the source can use to pass data when available. In the latter, the input port describes how the pure data product will read data from the sources (e.g., frequency, queries, etc.).

- **Output ports**: These specify the data assets exposed by the pure data product and the APIs through which consumers can access them. Through output ports, a pure data product can control how exposed data is consumed, limiting direct access to the underlying storage. This way, implementation choices can be hidden, maintaining a higher level of decoupling with consumers. Typically, a pure data product exposes the same data asset through multiple output ports and formats to enhance reusability and composability. For instance, one output port may expose user profile changes as events consumable in near real time, while another output port may expose the current state of all user profiles.

- **Discoverability ports**: These specify metadata exposed by the pure data product and the APIs through which consumers can access them. Some discoverability ports are common to all products, and their APIs are defined centrally to ensure greater interoperability. For example, every pure data product should have a discoverability port that, through a standard API, provides access to the product's descriptor document.

- **Control ports**: These specify configurations and commands that can be passed to the pure data product to modify its runtime behavior. APIs for passing information and commands should be standardized centrally and reused by every pure data product. Configurations (e.g., data refresh frequency, log level, etc.) and valid commands (activate, deactivate, etc.) may vary based on the specific product and should be specified in the descriptor document and exposed through a discoverability port. Control ports are generally not accessible to consumers. Because control ports are related to the operational management of a pure data product, they are usually accessible only by the product team and by platform-level tools used to automate product life cycle management activities.

- **Observability ports**: These specify operational data of a pure data product externally exposed (e.g., logs, quality metrics, service level indicators, etc.) and how they can be accessed. APIs for these ports should also be standardized centrally and adopted by all products to facilitate operational management.

All ports should be described in a machine-readable manner through APIs. The APIs used should abstract as much as possible from the technology used to implement them. Commonly used open specifications for describing APIs exposed by a port include OpenAPI, gRPC, SOAP, AsyncAPI, DataStoreAPI, and GraphQL.

In addition to APIs, each port must define the guarantees offered regarding delivered services (*promises*) and constraints on consumption methods (*expectations*). Typical promises include SLA, SLO, and deprecation policies, while common expectations include terms of use, query modalities, and payment policies. We'll delve deeper into defining and implementing the various ports of a pure data product in the upcoming chapters.

> **Hexagonal architecture**
>
> The hexagonal architecture is an architectural pattern proposed by Alistair Cockburn in 2005. Its goal is to break down an application into a set of loosely coupled modules, typically represented by hexagons. Each module communicates with other external systems through interfaces called ports. The number and type of ports can be defined as needed based on requirements. The fact that a module is represented by a hexagon does not constrain the number or type of ports to six. The key is to define the ports in a way that isolates the internal business logic of the module from the external world, avoiding unwanted dependencies between layers and contamination between implementation and the interface of exposed services. Pure data products draw inspiration from this pattern, sharing its goals, the way of graphically representing modules through hexagons, and the use of the term *ports* to define different interfaces.

Depending on their features and their role in the data value chain, pure data products can be categorized in various ways. In the upcoming section, we'll delve into some key classification methods that will be useful throughout the rest of the book.

Classifying pure data products

Pure data products can be classified in various ways depending on their characteristics and role within the architecture. In this section, we will explore some of the most popular classifications.

Source-aligned versus consumer-aligned

One of the most popular classifications divides pure data products based on their role within the architecture into the following three categories: source aligned, aggregation, and consumer aligned.

Source-aligned pure data products read data from external systems and transform them at the representation level to make data easily consumable by other products. They do not apply complex business logic; transformations are mostly syntactic. The semantics remain aligned with that of the source system from which they collect data. This type of pure data product generally reads data from a single source system. Any connections with data from other systems would require applying business logic. Consolidation operations are then performed as needed by other downstream products. The same source system may have multiple associated source-aligned products. Each pure data product associated with the same source system reads and makes different data assets available. The more complex a source system is, the more data assets it manages. As a consequence, the number of associated source-aligned products is greater. For example, it makes little sense to have a single source-aligned product associated with a CRM. More likely, there will be several products, each associated with a different data asset managed by the CRM (e.g., leads, opportunities, contracts, etc.).

Aggregation pure data products read data from other pure data products and connect them before redistributing them. Before redistributing, they may aggregate, filter, or normalize the data. It is generally not recommended to use this type of pure data product solely for connecting, filtering, or aggregating data owned by other data products. Linking, filtering, or aggregation operations should be performed directly by consumers with the support of a data virtualization layer. For example, if there is a need to provide a view that combines customer data collected from different source-aligned products, consumers should build this view in a self-service manner on the data virtualization layer. Implementing a specific pure data product for this task means creating copies of data and, at the same time, risks of inconsistency. The same applies if there is a need to filter the data from a source product or group it to create aggregated metrics. If the filter or aggregated metrics calculation logic is particularly complex, consumer-aligned pure data products can be built to redistribute the metrics or classification labels on which to filter the data at query time.

The only case where it makes sense to use this type of pure data product is to perform semantic normalization of data coming from different sources. The data itself is not modified or enriched. The pure data product, however, changes its representation by unifying the different semantics applied to data by the different sources. As we will see in *Chapter 10, Distributed Data Modeling*, the evolution of data virtualization systems in advanced semantic layers makes the creation of such pure data products increasingly unnecessary even in this case. The semantic layer can perform much of the semantic normalization between data exposed by different products for the consumer on the fly without requiring the creation of an ad hoc aggregation pure data product.

Consumer-aligned pure data products are all products that read data only from other products and just share the generated data they own. They are the products that enrich data to make them usable for the implementation of new use cases. Alignment with a specific use case may vary depending on the product. Some can be very generic and enable an unlimited range of use cases. Others may be more specific. In general, a consumer-aligned data product should always promote the reuse of the data assets it exposes and should not be tailored to the specific needs of a single use case.

Domain-aligned versus value stream-aligned

Pure data products can be grouped by the business domain they belong to. Business domains are typically mapped to business functions (e.g., marketing, sales, HR, etc.). Most pure data products only read data produced within one domain. Data acquired through input ports comes exclusively from systems or pure data products internal to the domain. These pure data products are of the **domain-aligned** type. In the early stages of adopting data-product-centric architectures, domain-aligned products are among the first to be developed. As the adoption process matures, there is a growing need to develop pure data products that cross-join data coming from different domains. Unlike domain-aligned pure data products, these pure data products do not aim to support domain-specific use cases. Instead, they focus on use cases of global interest, linked to corporate processes that span across domains (e.g., procure to pay, order to cash, customer management, etc.). We refer to these pure data products as **value stream aligned**. Pure data products of this type are more complex and present challenges related to attribution of ownership and management of the various semantics associated with cross-domain data they must handle. We will revisit and delve deeper into these issues in the third part of the book.

Other classifications

The classifications seen so far are not mutually exclusive. Other classifications can be defined as well in a way that is functional to the specific context. For example, pure data products can be classified based on maturity, size, risk level in relation to managed data assets, and so on.

It is common to define a meta-classification of pure data products into tiers, each associated with different governance and security policies. Tiers can be defined either by leveraging other classifications or with ad hoc rules. The tiers are ordered in levels. The higher the tier level to which a pure data product belongs, the greater and more stringent the global policies it must adhere to. Depending on how the tiers are defined, it may or may not be possible for a pure data product to change tier level during its life cycle.

Summary

In the professional data management community, there is currently no clear convergence on a shared definition of what a data product is. Often, existing definitions are very generic and not usable in practice.

In this chapter, we attempted to provide a clear and truly actionable definition of a data product. Specifically, we introduced pure data products to identify those data products that, unlike analytical applications, are solely aimed at facilitating the reuse and composability of data assets strategically for the organization. Pure data products serve as units of modularization in the data architecture, allowing the potential value of managed data to increase by making it available for the implementation of multiple use cases.

We have delved into what it means for a pure data product to be reusable and composable in detail. To these fundamental characteristics, we added relevance and accuracy, illustrating how these factors also have a decisive impact on the value of the data asset managed by pure data products.

Finally, we have examined the key components of a pure data product, which, in addition to managed data assets, include metadata, applications, infrastructure resources, and interfaces. A pure data product encompasses all the components necessary for deployment and autonomous management throughout its life cycle.

In the upcoming chapter, we'll expand our perspective, transitioning from the intricacies of individual pure data products to a higher-level analysis of the overarching architecture they belong to, along with its key characteristics and some examples.

Further reading

For more information on the topics covered in this chapter, please see the following resources:

- *Data Jujitsu: The Art of Turning Data into Product by DJ Patil (2012)*: `https://www.amazon.com/Data-Jujitsu-Turning-into-Product-ebook/dp/B008HMN5BE/`

- *The evolution of data products: The data that drives products is shifting from overt to covert by Mike Loukides (2011)*: `http://radar.oreilly.com/2011/09/evolution-of-data-products.html`

- *The Data Product ABCs by Juan Sequeda (2022)*: `https://www.datasciencecentral.com/data-product-framework/`

- *Measuring The Value Of Information: An Asset Valuation Approach by Daniel Moody, Peter Walsh (1999)*: `https://www.semanticscholar.org/paper/Measuring-the-Value-Of-Information-An-Asset-Moody-Walsh/bc8ee8f7e8509db17e85f8108d41ef3bed5f13cc`

- *Build data liquidity to accelerate data monetization by MIT Center for Information Systems Research (2021)*: `https://cisr.mit.edu/publication/2021_0501_DataLiquidity_WixomPiccoli`

- *The FAIR Guiding Principles for scientific data management and stewardship by Wilkinson, M., Dumontier, M., Aalbersberg, I. et al. (2016)*: `https://www.nature.com/articles/sdata201618`

- *Data Mesh: Data as a Product by Zhamak Dehghani (2022)*: `https://www.thoughtworks.com/about-us/events/webinars/core-principles-of-data-mesh/data-as-a-product`

- *Hexagonal architecture by Alistair Cockburn (2005)*: `https://alistair.cockburn.us/hexagonal-architecture/`

3

Data Product-Centered Architectures

In the previous chapter, we laid the groundwork by introducing the concept of a pure data product. We provided a formal definition and delved into its pivotal role in crafting sustainable and adaptable data management solutions.

In this chapter, we will delve into the principles that underpin a data product-centered architecture and its main components. Recognizing that a data management solution is a complex socio-technological system, we'll seamlessly blend our analysis of organizational architecture with that of technological architecture. To cap it off, we'll showcase alternative approaches to modern data management and examine how they architecturally align with the data product-centric approach outlined in this book.

This chapter will cover the following main topics:

- Designing a data-product-centric architecture
- Dissecting the architecture's operational plane
- Dissecting the architecture's management plane
- Exploring alternative approaches to modern data management

Designing a data-product-centric architecture

The transition from a monolithic data management solution to a modular one based on treating data as a product represents a true paradigm shift, demanding interventions at both the technological and organizational architecture levels. In this section, we'll explore the key elements to consider when designing such a socio-technological architecture.

System architecture

A **system** is a collection of interconnected components or elements that work together to achieve a common purpose or function. In our case, the system we aim to define the architecture for is a modular data management solution designed to make data assets easily usable in supporting the achievement of the organization's strategic objectives. But what exactly is the architecture of a system? Numerous definitions exist in the literature. Ralph Johnson, co-author of *Design Patterns: Elements of Reusable Object-Oriented Software*, defines the architecture of an IT system as "all the important stuff, whatever it is". The definition is certainly provocative but not trivial. The key idea is that everything playing a crucial role in determining the behavior and characteristics of a system is part of its architecture.

Another popular definition of architecture is provided by ISO/IEC/IEEE 42010:2011 and later adopted also by **The Open Group Architecture Framework** (**TOGAF**). In this case, architecture is defined as follows:

The fundamental concepts or properties of a system in its environment embodied in its elements, relationships, and in the principles of its design and evolution.

According to this definition, architecture isn't just the structure that holds a system's elements together, but also the processes that allow that structure to evolve. Thus, as shown in the following figure (*Figure 3.1*), we can divide a system's architecture into two parts: the core architecture and the meta-architecture.

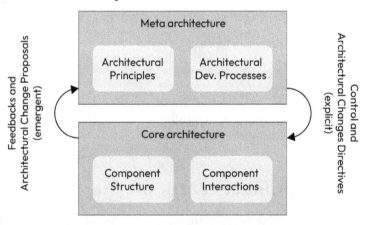

Figure 3.1 – System architecture

The **core architecture** encompasses the structure that connects the system's components and dictates their interactions with each other and the external environment. Conversely, the **meta-architecture** is comprised of the principles and processes that govern the development and evolution of the core architecture.

Every system has its architecture. In the case of **human-built systems**, such as a data management solution, the architecture can be intentional or emergent. It is intentional when explicitly designed at the outset to guide the development of individual components, while it is emergent when it autonomously arises from the interaction of developed components.

To build adaptive systems, those that need to evolve over time in response to changes in their operating environment, the meta-architecture is just as, if not more, important than the core architecture. In an adaptive system, a robust architecture should not only regulate the system's functioning (core architecture) but also its ability to evolve over time (meta-architecture). Explicitly defining the architecture is useful to influence the system's behavior, resulting from the interaction of its parts, and mitigate the risks of not achieving desired objectives (e.g., maximizing the value of managed data) or desired characteristics (e.g., sustainability and scalability). When the risk of system failure is high, as in the case of a data management solution, explicitly defining the system's architecture becomes necessary.

However, architecture imposes constraints on the autonomy of individual parts, centralizes certain functions, and adds communication overhead. In other words, it reduces the system's agility in favor of control to ensure the necessary cohesion among parts for goal attainment. Nonetheless, reduced agility in an uncertain and rapidly evolving context reduces the system's adaptability to sudden changes. Therefore, architectural design must consider these aspects and find the right trade-offs, based on the environment in which the system will operate, in terms of part autonomy and cohesion, agility and control, and centralization and decentralization.

The meta-architecture, which regulates the evolutionary capability of the architecture, becomes extremely important in this regard. It is the part to focus on predominantly at the beginning of the design activities. Having a framework to guide architectural developments enables the possibility of defining only the parts of the core architecture needed to start the solution implementation, extending them later only if useful or necessary. Developing and evolving the architecture concurrently with the solution's development and evolution, as depicted in the following figure (*Figure 3.2*), allows deferring architectural decisions until they are truly necessary (**last responsible moment**), or, in other words, when the risk of not deciding outweighs the risk of making a wrong decision.

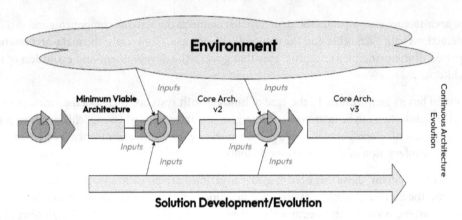

Figure 3.2 – Continuous architecture evolution

Delaying decisions allows for the acquisition of more information about the system and greater architectural maturity. This, in turn, eases both the decision-making process and the subsequent adoption of architectural decisions. It also grants more autonomy to the system, enabling architecture components to emerge spontaneously as a consequence of its evolution. These emergent architectural components can be then explicitly integrated or rejected into the overall architecture, following the processes defined at the meta-architecture level.

The architecture of the data management solution is part of the broader architecture of the organization. Each organization is different in terms of mission, culture, history, and operating environment. Therefore, the architecture must be designed and evolved based on the specific context in which the solution will be developed and maintained. Simply copying architectures successfully used in other contexts generally does not work (**cargo cult thinking**).

In *Chapter 8*, *Moving through the Adoption Journey*, and *Chapter 9*, *Team Topologies and Data Ownership at Scale*, we will explore how to define and develop a product-centric architecture that is tailor-made to a given specific context.

In the rest of this chapter, we will instead examine the key characteristics of this type of architecture and its main components.

Socio-technical architecture

An **organization** is a complex system consisting of people and technologies. A **data management solution** is a subsystem of the organization. It is also a complex system composed of people and technologies. Systems of this kind, systems composed of people and technologies, are called **socio-technical systems**.

The architecture of a socio-technical system is divided into **social architecture** and **technological architecture**. Social architecture defines how people are structured into functional units and how these functional units interact to generate desired outcomes. Similarly, technological architecture defines how technologies are structured into functional units and how these functional units interact to generate desired outcomes.

In a socio-technical system, there are strong correlations between people and technologies, as well as between social architecture and technological architecture, as shown in the following figure (*Figure 3.3*).

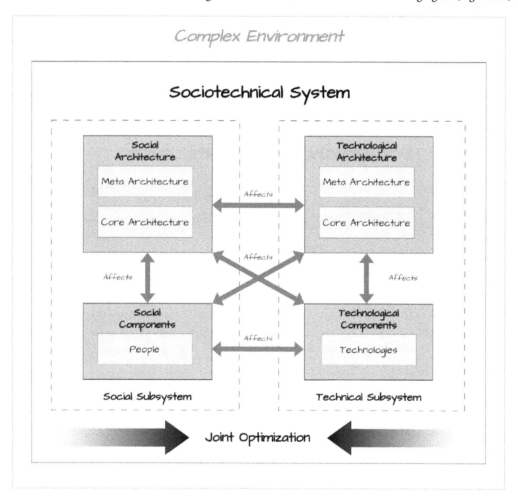

Figure 3.3 – Socio-technical system

Since, in a socio-technical system, people and technology are strongly interrelated, it is necessary to design **social architecture** in synergy with **technological architecture**, seeking a joint optimization that allows the system to achieve its goals in a sustainable way over time.

Designing the two architectures separately is risky. If there is no alignment between social architecture and technological architecture, generally, social architecture prevails and ends up determining technological architecture (**Conway's law**). Trying to derive one architecture from the other is also risky. If one tries to design organizational architecture based on technological architecture (**reverse Conway maneuver**), the system risks failing due to the passive resistance stemming from the natural inertia of the organization to change. Conversely, if one tries to design technological architecture based on organizational architecture, the system risks missing out on opportunities provided by technological innovations and losing competitiveness. That's why the two architectures need to be developed in tandem, incrementally and synergistically. In other words, the process of designing, governing, and evolving social and technological architecture must be unified, as shown in the following figure (*Figure 3.4*).

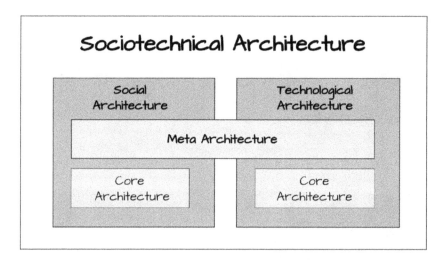

Figure 3.4 – Sociotechnical system architecture

Architectural principles

Architectural principles are general rules and guidelines designed to endure over time and be rarely modified. They guide, support, and harmonize architectural choices. Architectural principles determine how the architecture of a system is defined, governed, and evolved to ensure that the system can achieve its goals in the desired manner.

A data management solution centered around data products is based on three key architectural principles: modularity, composability, and sustainability. Let's delve into each of them in more detail, examining their rationale and implications:

- **Modularity**: The architecture must be modular. Each module should be self-contained, loosely coupled with other modules, and capable of autonomously producing tangible value for the system:

  - **Rationale**: Modularization aims to divide the complexity of the system into parts that are easier to manage, considering the natural limits of teams' cognitive capacity.

  - **Implications**: The technological architecture's unit of modularization is the data product. The organizational architecture's unit of modularization is the team. A **modular architecture** is generally partitioned by functions (e.g., governance, data product development, platform engineering, etc.) rather than divided by layers (e.g., ingestion, processing, analysis, etc.). A **socio-technical modular architecture** can be distributed, centralized, or hybrid. The technological architecture is completely centralized when all teams use the same technological stack, and completely decentralized when each team is free to choose which technologies to use. Social architecture is completely centralized when all functions necessary for system development are concentrated in a single team or organizational function, and completely decentralized when each function is associated with one or more teams or organizational functions. The architecture is hybrid if it is neither completely centralized nor completely decentralized.

- **Composability**: The architecture must facilitate the composability of its modules:

  - **Rationale**: Composability aims to allow the reuse of different autonomous modules that make up the architecture in multiple configurations to support different functions and objectives over time.

  - **Implications**: The technological architecture is composable when individual modules are interoperable, meaning communication protocols and interaction modes are encoded through shared and globally respected specifications. Social architecture is composable when it can integrate, build, and reconfigure internal and external skills to respond to environmental changes (**dynamic capabilities**). In general, to ensure composability at both technological and organizational levels, it is preferable to manage the system's deliverables, whether they are parts of the solution or the organizational model, as products rather than projects.

- **Sustainability**: The architecture must allow sustainable evolution of the system over time within a complex and constantly changing environment:

  - **Rationale**: Architectural modularity and composability allow the system to quickly adapt to contextual changes. However, to ensure the system's survival over time, it is important to ensure that its evolution occurs sustainably.

- **Implications**: The technological architecture is sustainable when it ensures constant levels of quality, security, and ease of system management. Generally, these characteristics are guaranteed through a self-serve platform capable of automating security/quality checks and operational processes. Social architecture is sustainable when it allows planning and prioritizing activities based on a balanced set of indicators that monitor both the system's behavior in relation to declared strategic objectives (metric of success) and its overall health (fitness functions).

Architectural components

We've seen the key principles of a socio-technological architecture centered around data products. Let's now delve into the main components of the system we aim to design – a modern data management solution.

A socio-technical system, such as an organization, consists of logical functions. Each logical function has specific capabilities. A **capability** describes what the system can do. An **organizational function** (e.g., business unit, department, etc.) employs people and technologies to instantiate a logical function. In other words, an organizational function linked to a capability specifies how the capability is concretely implemented using the resources of the system (people and technologies). Capabilities are the means through which the organization realizes and supports the value streams necessary to implement its business model.

Logical functions serve as socio-technical subsystems within the broader organizational system. We use the **viable system model** (VSM) described by Stafford Beer in the book titled *Brain of the Firm* to identify the key logical functions that an organization capable of evolving over time in a changing environment must have. A **viable system**, as shown in the following figure (*Figure 3.5*), consists of five logical functions or subsystems, where the first one (System 1) has operational capabilities, while the remaining four (Systems 2-5) have coordination and management capabilities.

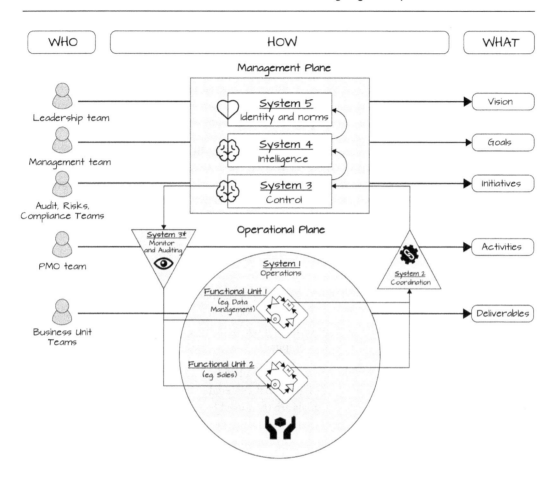

Figure 3.5 – VSM

System 1 (operations system) oversees the day-to-day activities critical to the organization, ensuring smooth operations and sustained efficiency. System 1 is a set of functions or subsystems generally mapped to the organizational units or business domains of the organization (e.g., HR, marketing, sales, etc.).

System 2 (coordination system) handles operational coordination, ensuring that the activities performed by System 1 are integrated and coherent. It acts as a bridge between the various functional units of System 1, facilitating communication and collaboration. It also provides key information on the progress of operations to the control system (System 3). The coordination system plays a crucial role in maintaining organizational unity and preventing system fragmentation. Within an organization, System 2 can be mapped, for example, to the **project management office (PMO)** team.

System 3 (control system) deals with operational planning, ensuring that operational activities are aligned with the organization's goals (strategy). The control system plays a crucial role in defining operational priorities, which are derived from strategic objectives. It allocates the required resources for the execution of prioritized activities and ensures that these activities adhere to the organization's policies. The audit and monitoring function of operational activities can be mapped onto an independent system closely linked to System 3, generally referred to as System 3*. Finally, the control system utilizes information on results obtained from the coordination system (System 2) to initiate a new cycle of prioritization.

The control system acts as a regulating system, identifying deviations and initiating corrective actions at the operational or strategic level to bring the organization back on track if necessary. The control system is crucial for maintaining stability and adaptability. Within an organization, System 3 can be mapped, for example, to the controller, audit, compliance, and risk teams.

System 4 (intelligence system) is responsible for strategic planning, translating the vision formalized by the identity system (System 5) into a set of goals that can enable the organization to implement its vision within the operating context. The goals are then used by the control system (System 3) to define, prioritize, and evaluate the results of operational activities. The intelligence system is crucial for ensuring long-term sustainability. Within an organization, System 4 can be mapped, for example, to the management team.

System 5 (identity system) defines the values, purpose, and evolutionary directions of the organization. It sets long-term goals that guide the organization's behaviors and define its identity. The vision defined by the identity system is then translated into a strategic plan by the intelligence system (System 4). Within an organization, System 5 can be mapped, for example, to the leadership team.

The functions that make up the operations system (System 1) are viable systems themself. Each function is composed of sub-functions that perform operational activities (System 1) and a set of systems (System 2-5) that manage the operation of the specific sub-function.

Let's take a look at *Figure 3.6*, which illustrates how the VSM applies to the **data management function**.

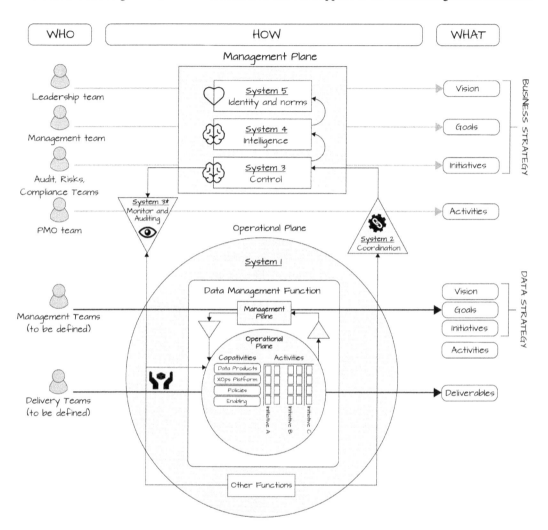

Figure 3.6 – Data management function

A **data management function** is one of the sub-functions of the operations system (System 1) of the organization. It is a viable system whose external environment is the organization itself. Defining the socio-technical architecture of a data management solution means defining how its operational and control subsystems are structured and how they interact with each other and with the rest of the organization to achieve the goal of making relevant, accurate, reusable, and composable data products available to other functions. Depending on the architecture, the data management function can be associated with a single organizational unit or distributed across multiple organizational units of the organization.

In the following two sections, we will delve into the roles of operational subsystems and management subsystems in the context of a data management solution (*Figure 3.6*) to lay the foundations for its architecture definition.

Dissecting the architecture's operational plane

The goal of the data management function is to create a data management solution capable of supporting other functions and the entire organization in achieving strategic objectives. To develop a modular and data-product-centric data management solution, the data management function must possess, at the operational level (System 1), specific core capabilities. In this section, we will explore what these capabilities are and why they are necessary to create a modular, composable, and sustainable data management solution.

Core capabilities

The four core capabilities that the data management function must have to build a product-centric data management solution are data product development, governance policy-making, **extended operations** (**XOps**) platform engineering, and data transformation enabling (*Figure 3.7*).

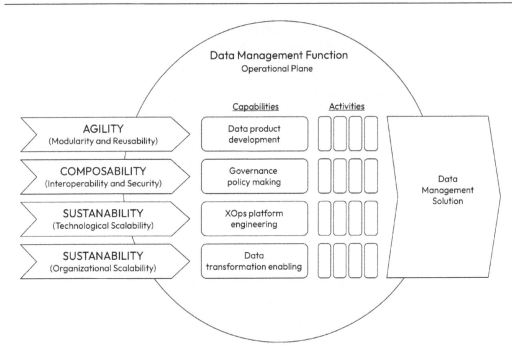

Figure 3.7 – Data management function's core capabilities

The capability to develop and evolve data products over time is obviously necessary to build modular and agile solutions, where data is managed as a product. However, it alone is not sufficient to ensure the desired characteristics of composability and sustainability. To this end, three additional capabilities must be added:

- The capability to develop and manage global governance policies to regulate the development and interaction modes between pure data products is necessary to ensure interoperability and security (composability).

- The capability to develop and manage an XOps platform supporting product development and management is necessary to ensure the scalability required for the system to grow sustainably.

- The capability to support the socialization process of architecture and organizational learning (data transformation enabling) is necessary to ensure the organic growth of the system.

How these capabilities map to the organizational functions is an architectural choice. When all capabilities are mapped to the same organizational function, it encompasses the entire data management function. Regardless of the architecture, it is important to manage budgets supporting individual capabilities separately. Having a single, undistinguished budget associated with the data management function does not allow for a clear and transparent balance of resources invested in various deliverables.

Typically, there's a risk that the development of data products consumes a significant portion of the available budget, resulting in a shortage of deliverables from other capabilities essential for ensuring the desired composability and evolvability of the overall data management solution.

Data product development

The capability to develop and manage data products is the main capability of a data management function with the goal of building a modular data management solution. Other capabilities support it to ensure the proper balance of the released solution in terms of composability and sustainability.

We have extensively discussed in the previous chapters what it means to manage data as a product. Now, let's explore some of the main organizational and technical architectural choices to be made in relation to this capability of the data management function:

- **Technical architecture**:

  - What are the minimum characteristics that a pure data product must have to be considered as such? What is necessary for it to be released?

  - Which specification should be used to describe a pure data product or at least its public interfaces (data contract)? What minimal information should it contain?

  - Is there a reference technology stack for the implementation of pure data products, or can each product freely choose its own?

- **Organizational architecture**:

  - How is the team or teams that will be responsible for the development of data products composed? How are they mapped in relation to organizational units?

  - What are the key roles within the team? Who is accountable, who is responsible?

We will delve deeper into these themes related to the development of data products in the second part of the book (*How to Manage the Data Product Life Cycle*).

Governance policy-making

The capability to define governance policies is a supporting capability. Its main purpose is to ensure that the pure data products released are safely composable. Policies defined at the data management function level add to those defined at higher levels of the organizational architecture. The type of policies that can be defined at this level, and their specificity, determine the extent to which one wants to promote integration and standardization (broad scope) versus agility and diversification (narrow scope) in the developed data management solution. *Figure 3.8* illustrates the main types of policies that can be defined by the data management function.

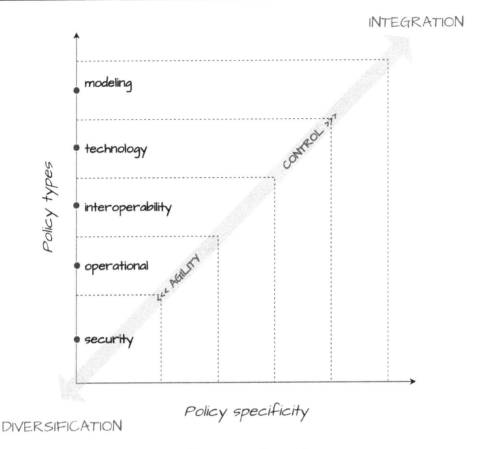

Figure 3.8 – Governance policy-making scope

The first three types of policies (security, operational, and interoperability) are foundational. They can be more or less specific but are essential. The data management solution must ensure a minimum level of **security**, **operational reliability**, and **interoperability** among products. Policies related to data technologies and modeling methods are optional.

The governance policy production capability does not include the ability to enforce them. **Enforcement** is part of the control function (System 3). Generally, in modern data management solutions, policy enforcement is automated as much as possible, with the control function performing just secondary-level verification. To automate enforcement, policies must be defined in a machine-readable format (**computational policy**). The XOps platform, described in the next section, is responsible for providing the necessary services to automate the enforcement of various computational policies.

It is advisable to implement the capability to define governance policies so that they are managed as products throughout their natural life cycle.

The following are some of the main organizational and technical architectural choices to be made regarding this capability of the data management function:

- **Technical architecture**:

  - What is the life cycle of a policy?

  - Which tools should be used to support the development and sharing of policies?

  - What specifications should be used to describe policies explicitly, accurately, and preferably in a machine-readable format?

  - How are policies monitored, and by whom?

- **Organizational architecture**:

  - How is the team or teams responsible for policy development structured? How are they mapped to organizational units?

  - What roles contribute to the definition and evolution of a policy? Who is accountable, and who is responsible?

  - What is the decision-making process leading to the definition, modification, or withdrawal of a policy?

We'll delve deeper into these policy-related topics in the second part of the book (*How to Manage the Data Product Life Cycle*). Data modeling policies will be further explored in *Chapter 10, Distributed Data Modeling*.

XOps platform engineering

The capability of **XOps platform engineering** is a supporting capability. Its primary purpose is to facilitate the management of pure data products throughout their life cycle and ensure their compliance with governance policies (*Figure 3.9*).

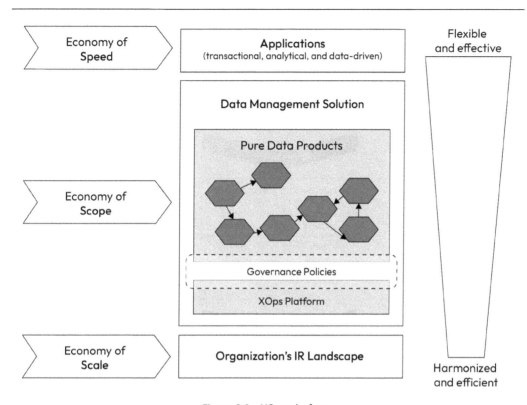

Figure 3.9 – XOps platform

 The deliverables of this capability contribute to the construction of an **XOps platform** that consolidates all the common functionalities necessary in the development and management of pure data products, such as publication, validation, monitoring, cost control, and search, among others. All the functionalities of the XOps platform are exposed to pure data products and, more broadly, to all users of the data management solution through self-serve accessible services. The exposed services are always in support of operations and compliance, meaning they do not implement specific business logic. The implementation of business logic is exclusively delegated to pure data products.

Observe in *Figure 3.10* how the services exposed by an XOps platform, regardless of its detailed architecture and specific implemented functionalities, can be grouped into two main planes: the **utility plane** and the **product control plane**.

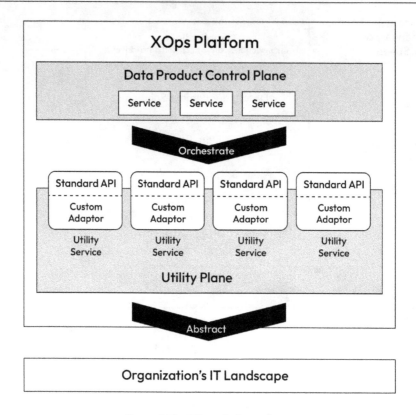

Figure 3.10 – XOps platform planes

The services of the utility plane provide an abstraction layer above the services provided by the underlying applications and platforms that make up the specific existing IT landscape. They aim to simplify access to these services by reducing their generality to functionally meet the specific needs arising from the actual usage requirements within the implemented data management solution. Additionally, they enable the decoupling of developments in the data management solution from the underlying technologies and platforms. For example, a service of the utility plane could provide an interface for tracking logs generated by various data products. This interface could be implemented through an adapter that stores logs in a commercially available product already present in the organization's IT landscape. If there is a future decision to change the log management product, it would suffice to implement a new adapter conforming to the interface exposed by the utility plane for log tracking. Existing data products would not be impacted by the change in the underlying log management product.

The services of the **data product control plane** orchestrate the services offered by the utility plane to provide higher-level functional capabilities for managing pure data products throughout their life cycle. The scope of functionalities common to this plane depends on how much governance policies push the system toward integration versus diversification. If there are few shared governance rules, the system tends toward diversification, and consequently, only a few functions can truly be standardized and consolidated through platform services. The platform is then reduced to almost exclusively the utility plane, providing access to basic infrastructure and application resources. Conversely, if governance rules are numerous and drive strong integration between system parts, many functions can be standardized and implemented through platform services.

 Thanks to the XOps platform, efficiency and compliance can be achieved, allowing the entire system to have the necessary scalability for sustainable growth.

Gregor Hohpe, in the book titled *Platform Strategy*, lists the main advantages of a platform, such as the XOps platform described here. This is summarized in *Table 3.1*.

Business Objective	Techinical Motivation	Implementation
Minimize mistakes	Hide complexity, support composability	Dafault settings, templates
Increase velocity	Reduce friction	Automation
Improve products	Fill product gaps	New components (e.g., product sidecar)
Enforce compliance	Restrict choice	Controls or wrappers
Reduce lock-in	Shield complexity	Wrappers / abstraction layers

Table 3.1 – Advantages of the platform

It's advisable to implement the XOps platform engineering capability so that the developed platform is managed as a product or a collection of products.

The following are some of the main organizational and technical architectural choices to be made concerning this capability of the data management function:

- **Technical architecture**:

  - When should one start implementing an XOps platform? Before developing products or at a later stage?

  - What are the essential functionalities that constitute the thinnest viable platform?

  - Should the platform be implemented with make or buy logic?

  - If make logic is chosen, what should be the architecture of the platform? Monolithic, layered, based on microservices, or other?

- **Organizational architecture**:

  - Is the use of the platform mandatory or optional?

  - How is the team or teams responsible for platform development structured? How are they mapped to organizational units?

  - What roles contribute to the definition and evolution of the platform? Who is accountable, who is responsible?

 We'll delve deeper into these topics related to the XOps platform in *Chapter 7, Automating Data Product Life Cycle Management.*

Data transformation enabling

The **data transformation enabling** capability is a supporting capability. Its primary purpose is to facilitate the dissemination of skills necessary for developing the data management solution according to the new paradigm. It's not a fleeting capability, useful only at the start of activities. The data management solution is designed to evolve over time, and its socio-technical architecture is constantly changing to adapt to the external environment and lessons learned during system development. The social component of the architecture has greater inertia to change compared to the technological one. For this reason, the data transformation enabling capability, like other core capabilities, is necessary as long as the system exists, or, in other words, until the data management solution is decommissioned.

The enabling capability does not engage in exploring new practices or skills, nor does it translate what is learned into organizational practices. Its task is to disseminate knowledge and practices already acquired among the organizational units to improve the effectiveness and efficiency of their work. Its main role is to balance organizational inertia to change by promoting the rapid spread of skills.

The following are some of the main organizational and technical architectural choices to be made concerning this capability of the data management function:

- **Technical architecture**:

  - What are the tools or methodologies that can be used to facilitate enabling activities?

- **Organizational architecture**:

  - How are the teams responsible for enabling activities structured? How are they mapped to organizational units?

  - What roles contribute to the management of enabling activities? Who is accountable, who is responsible?

  - Does it make sense to consider leveraging external resources to accelerate enabling activities?

We'll delve deeper into these enabling-related topics in *Chapter 9, Team Topologies and Data Ownership at Scale*.

In the next section, we will delve into the structure of the management plane, its subsystems, and their responsibilities.

Dissecting the architecture's management plane

The operational capabilities of the data management function must be managed to ensure that the implemented solution aligns with the organization's global objectives and strategy. This is the responsibility of the management functions (System 2-5) that we will analyze in detail in this section.

Identity system

The **identity system (System 5)** within the data management function plays a crucial role in shaping the vision and foundational principles that underpin the data strategy. The vision and principles of the data management function must align with the vision and principles of the organization.

The *vision* provides a high-level description of the desired future state for the data capabilities of the organization. It offers a clear and challenging picture of how data will be leveraged to drive innovation, improve decision-making processes, and support the organization in achieving its goals. The vision guides the definition of the data strategy, inspiring stakeholders and aligning their efforts toward a common objective. The vision may be complemented by a set of *values* that reflect those of the organization, adapting them to the data management context (e.g., transparency, accountability, simplicity, innovation, continuous learning, passion, etc.). Values define the core beliefs that broadly guide priorities in implementing the vision.

Principles describe the fundamental guidelines or rules that guide decision-making processes and actions aimed at implementing the vision. Unlike values, principles are more concrete and immediately actionable. They directly refer to the construction of the socio-technical architecture of the solution. They are architectural principles. In the case of architectures centered on data products, key principles include modularity, composability, and evolvability. Additional principles may be added based on the specific context.

Typically, the person in charge of the data management function (e.g., CIO, CDO, data platform lead, etc.) collaborates with their direct reports to define the vision, values, and principles on which to base the data strategy. Vision, values, and principles tend to remain stable over time. However, they are not immutable, and periodically, usually on an annual basis, they can be reviewed, corrected, and integrated.

Intelligent system

The **intelligent system (System 4)** within the data management function plays a pivotal role in translating the vision outlined by the identity system into tangible strategic objectives, commonly referred to as goals. These goals articulate how the function aims to realize its vision, emphasizing desired outcomes rather than prescribing specific solutions, ideas, or features. Essentially, they answer the question of *what* needs to be achieved, leaving the *how* to be determined later.

To formulate goals, the intelligent system, in addition to having a clear understanding of the vision, values, and principles defined by the identity system, must gather input from the rest of the organization, the external environment, and the underlying control function.

Typically, the reports of the data management function manager, collaborating with their respective teams, define these goals. Regular reviews, corrections, and integrations, usually conducted every six months, ensure that the goals remain adaptive to evolving organizational priorities and external dynamics.

Control system

The **control system (System 3)** within the data management function is responsible for translating the goals defined by the intelligent system into an operational plan composed of specific initiatives. An initiative represents an operational development line that supports the achievement of a goal. Each goal may have several associated initiatives. It is preferable to define each initiative in alignment with a specific goal, avoiding initiatives associated with multiple goals. An example of an initiative associated with the goal of democratizing data access could be the development of a data catalog.

Each initiative consists of operational activities that will then be implemented through the capabilities of the operational system (System 1). For example, developing a data catalog initiative can be translated into multiple activities at the product level (e.g., adding data schema to the data contract), governance policy level (e.g., specifying a common format for representing data schemas), XOps platform level (e.g., developing an adapter to read schemas in the contract and publish them in the catalog), and enabling level (e.g., training potential business users on using the catalog).

The control function, in addition to defining initiatives, is tasked with prioritizing operational activities to facilitate their implementation, allocating the necessary resources for development, monitoring their results, and only intervening when necessary.

Typically, the control function collaborates with the intelligence function to define initiatives. It works with the coordination function to define the activities within each initiative and their priority. Initiatives and their associated budgets can be modified, usually on a quarterly basis. The activities within an initiative and their priorities can be adjusted, typically on a monthly basis.

Coordination system

The coordination system (system 3) within the data management function is responsible for coordinating the operational activities associated with each initiative defined by the control system. It can be linked to an external PMO function separate from the operational teams, or, following the Scrum of Scrums model, to a federated team consisting of leaders from various operational teams. Coordination activities are usually reviewed on a weekly basis.

Operating model

Figure 3.11 shows different ways in which management functions interact with each other and the operational capabilities that constitute the operating model of the function, which is, in our case, the data management function.

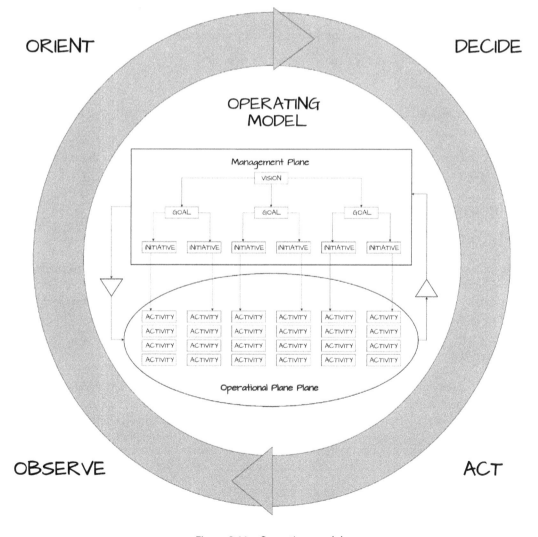

Figure 3.11 – Operating model

The definition of the operating model is a key component in designing the socio-technical architecture of the system.

The following are some of the main architectural choices to be made in defining the operating model:

- What roles are involved in defining the vision, goals, initiatives, and activities? Who is accountable, who is responsible?

- How often are vision, goals, initiatives, and activities reviewed?

- How is alignment and conformity ensured between the vision, values, and principles defined at the data management function level and those defined at the corporate level?

- How are the vision, values, and principles communicated within and outside the data management function?

- How are initiatives and activities prioritized?

- How is resource allocation managed for the development of individual activities?

- How are activities within the same initiative coordinated?

- How is the impact of initiatives on associated goals measured?

We'll delve deeper into these operational model-related themes in *Chapter 8, Moving through the Adoption Journey*).

In the next section, we will explore some modern approaches to data management describing the aspects of the socio-technological architecture seen so far.

Exploring alternative approaches to modern data management

The data management approach centered around the concept of data products is not the only modern approach to the practice aimed at addressing issues faced by traditional solutions. In this section, we will review the three main alternative approaches, examine their key characteristics, and explore how they relate to the data-product-centric approach described in this book.

Data mesh

Data mesh is a decentralized socio-technical approach to data management for analytical purposes in large and complex organizations. Coined by Zhamak Dehghani, the data mesh approach is based on the following four principles that define its technological and organizational architecture:

- **Domain ownership**: Data ownership is decentralized to business domains, which are responsible for selecting, managing, governing, and sharing them

- **Data as a product**: Data under the responsibility of business domains must be managed as a product to enable reuse across various domains, avoiding the formation of different data silos

- **Self-serve data platform**: A self-service platform should be in place to support product teams in development, reducing overall ownership costs

- **Federated computational governance**: A federated data governance team composed of domain leaders, product owners, platform owners, enterprise architects, and other subject matter experts must exist to define global policies necessary to balance domain autonomy in favor of mesh interoperability

Figure 3.12 illustrates how the principles of the data mesh are interconnected.

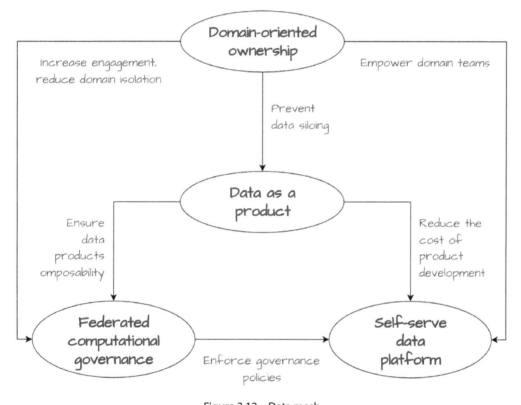

Figure 3.12 – Data mesh

The data mesh approach shares many commonalities with the one presented in this book. Both are built on the idea of managing data as products. Moreover, both identify global governance policies and self-service platforms as crucial elements for ensuring the sustainability of the overall solution. However, there are also significant differences in terms of scope and organizational model.

In terms of scope, the data mesh aims to build scalable data management solutions to support specifically analytical applications. The data-product-centric approach presented in this book focuses on managing data collected through process digitalization for the construction of various types of applications, both analytical and non-analytical.

Regarding the operational model, the data mesh adopts a strongly decentralized approach. Data responsibility is distributed across various business domains, while governance is managed by a federated team that also serves as the sole control and coordination function. Only the platform development capability is implemented through a centralized team. The approach described in this book is less prescriptive in terms of the operational model. The modularization of the solution through data products allows for decentralization, moving toward a mesh approach, but it is not mandatory. No element of the proposed approach assumes a transition to a fully decentralized model.

The decentralization inherent in the data mesh approach introduces an element that requires not only a paradigm shift to embrace the management of data as a product but also a shift from a centralized to a completely decentralized model. The transition to a mesh approach, therefore, demands a strong commitment from individual business domains to reconfigure organizationally and create new internal data management capabilities. Even in organizations where such commitment exists, a common approach to adopting a data mesh is to start with a centralized data product-based architecture and then gradually and incrementally decentralize parts of the operational model. For other organizations, either the commitment is lacking or the need to scale data management solutions does not necessitate complete decentralization, making the data mesh approach either impractical or economically unfeasible.

Data fabric

Data fabric is an approach to data management that focuses on technological architecture to make data integration flexible, reusable, and automated, facilitating data access within organizations.

Coined by Gartner, the data fabric approach is based on the following three principles that define its technological architecture:

- **All data**: The data fabric must be able to handle all data produced by the organization, consolidating it according to different integration styles and leveraging existing technologies where possible.

- **All metadata**: In addition to managing data, the data fabric handles all associated metadata. This includes technical metadata related to origin and transformation methods, as well as business metadata related to semantics and data usage.

- **Intelligent automation**: The data fabric should represent data and metadata through a knowledge graph. The knowledge graph serves as the foundation for applying AI techniques to automate data management processes.

Figure 3.13 gives a high-level view of the technological architecture of data fabric.

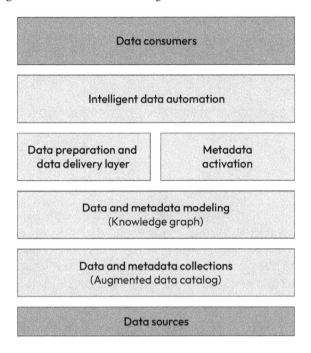

Figure 3.13 – Data fabric

The data fabric approach focuses more on the integration platform than on how the integrated data is structured and managed. The concept of a data product is not part of its definition. The solutions built on top of the data fabric can vary in modularity. However, it is evident that to construct a data fabric, a high level of standardization in building these solutions is crucial. In the absence of uniformity, it is challenging to consolidate functions to automate integration processes. Another central element for building a data fabric is the completeness of the metadata that can be gathered. Without proactive and comprehensive management of all relevant metadata, enabling their activation through intelligent automation techniques is not feasible. For these reasons, a data management solution based on data products with a high level of integration (i.e., broad scope and specificity of global governance policies) constitutes a fertile foundation for the development of a data fabric. In this context, the data fabric can be seen as a natural evolution of the XOps platform.

Data-centric approach and dataware

The **data-centric approach** is a model-driven approach to data management. It is based on the idea of building a single centralized data model that supports the needs of various applications. Core business rules are contained within the model, not within the applications. Applications read and write data from the central model in accordance with its structure and the rules governing its development (*Figure 3.14*).

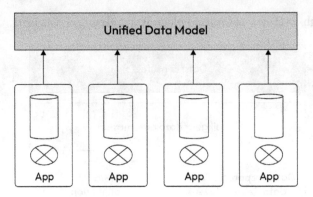

Figure 3.14 – Data-centric approach

Generally, the central data model is defined through a **knowledge graph** that connects the abstract model, also known as **ontology** (i.e., abstract concepts and relationships between concepts), with **factual data** (i.e., instances of concepts) present in storage systems. The knowledge graph serves as a centralized semantic layer through which all applications, both transactional and analytical, access the data they need. Physical data management can be delegated to individual applications as long as the model remains compliant with the central ontology.

Dataware is the term used to reference the data-centric approach when, together with the data model, even the management of physical storage is centralized. *Figure 3.15* shows the dataware approach is a specific type of data-centric approach.

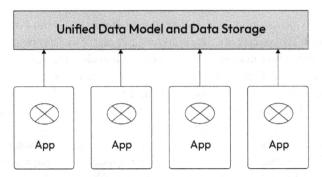

Figure 3.15 – Dataware approach

The data-centric approach, with or without the centralization of physical data management, can be layered on top of the data fabric approach. The data fabric aims to build a knowledge graph to automate data integration and consumption processes. The construction of the knowledge graph is generally done ex-post to reconcile fragmented data and metadata within the existing application landscape. However, it is not obligatory. The data fabric's knowledge graph can also be built ex-post for all existing applications while being built ex-ante, following a typical model-driven approach of data centrism, for all new applications. The data-centric approach, like the data fabric approach, does not dictate how to organize data. It only requires that the managed data conforms to the central data model. Therefore, it is possible to use the data-centric approach in conjunction with the product-centric approach.

Summary

In this chapter, we have seen how the architecture of a data management solution should be designed synergistically from both organizational and technological perspectives.

Since there is no one-size-fits-all architecture for data management solutions that can work in all contexts, in this chapter, we focused on high-level elements and architectural principles that typically characterize the architectures centered on data products.

We first explored the three key principles that should guide the development of a socio-technical architecture for a data management solution centered around data products: modularity, composability, and sustainability.

We have then used the VSM to describe the main operational and management functions that make up the architecture of such systems, listing the key architectural choices to be made for each one.

Finally, we introduced popular alternative approaches to data management: data mesh, data fabric, and data centrism/dataware. For each of them, we highlighted the main architectural characteristics and their relationships with the data-product-centric approach discussed in this book.

We will leverage this foundational architectural knowledge throughout the rest of the book to explain how to build a tailored-made data-product-centric architecture that fits within your specific context.

In the next chapter, we will begin the second part of the book dedicated to managing pure data products throughout their life cycles. Specifically, in the next chapter, we will start exploring how to identify relevant pure data products and prioritize their development.

Further reading

For more information on the topics covered in this chapter, please see the following resources:

- *Who Needs an Architect?, Martin Fowler (2003)*: `https://martinfowler.com/ieeeSoftware/whoNeedsArchitect.pdf`

- *Building Evolutionary Architectures, N. Ford, R. Parsons, P. Kua (2017)*: `https://www.amazon.it/Building-Evolutionary-Architectures-Support-Constant/dp/1491986360/`

- *Continuous Architecture in Practice, di M. Erder, E. Woods, P. Pureur (2021)*: `https://www.amazon.it/Continuous-Architecture-Practice-Software-Agility/dp/0136523560`

- *Soulful SocioTechnical Architecture, Marco Consolaro (2021)*: `https://www.youtube.com/watch?v=uGggRmBwKK0`

- *Viable System Model Introduction, Mark Lambertz (2021)*: `https://www.youtube.com/playlist?list=PLgCHVSvQeueVok0dbWIqOeBFOrTugsXmm`

- *The Fractal Organization, Patrick Hoverstadt (2009)*: `https://www.amazon.it/Fractal-Organization-Creating-Sustainable-Organizations/dp/0470060565`

- *The Magic of Platforms, Gregor Hohpe (2021)*: `https://www.youtube.com/watch?v=DgsIDqkvLME`

- *Data Mesh, Zhamak Dehghani (2022)*: `https://www.amazon.it/Data-Mesh-Delivering-Data-driven-Value/dp/1492092398/`

- *Principles of Data Fabric, Sonia Mezzetta (2023)*: `https://www.amazon.it/Principles-Data-Fabric-organization-implementing/dp/1804615226/`

- *The Data-Centric Revolution, Dave McComb (2022)*: `https://www.amazon.it/Data-Centric-Revolution-Restoring-Enterprise-Information/dp/1634625404`

- *What is Dataware?, Joe Hilleary (2022)*: `https://cinchy.com/blog/what-is-dataware`

Part 2: Managing the Data Product Lifecycle

In this part, we'll explore how to manage a data product throughout its lifecycle. We'll begin by examining how to identify the data products to develop, prioritize them using a business-case-driven approach, and design them based on collected requirements. We will then look at how to handle releases and data product management in a production environment by adapting common DevOps practices to the data context. Finally, we will explore how to automate data product lifecycle management through a self-serve platform, analyzing its core capabilities, architecture, and implementation options (make versus buy).

This part has the following chapters:

- *Chapter 4, Identifying Data Products and Prioritizing Developments*
- *Chapter 5, Designing and Implementing Data Products*
- *Chapter 6, Operating Data Products in Production*
- *Chapter 7, Automating Data Product Lifecycle Management*

4

Identifying Data Products and Prioritizing Developments

In the previous chapter, we examined the components and structure of the socio-technical architecture that underpins a data management solution centered on data products.

In this chapter, our focus shifts toward the identification and prioritization of data products to implement. We'll first introduce the domain-driven design methodology as a way to identify the relevant business cases to address within the business domain of an organization. Then, we'll explore how to utilize the event storming methodology to identify the necessary data products to support the implementation of the business cases of interest. Finally, we'll tackle the management of the data product portfolio. This involves validating the identified data products, describing them using a business model canvas, and ultimately, ensuring the ongoing coherence and efficacy of the product portfolio through regular monitoring of its structure.

This chapter will cover the following main topics:

- Modeling a business domain
- Discovering data products with event storming
- Managing a data product portfolio

Modeling a business domain

A pure data product provides a solution to a problem related to the management and sharing of one or more data assets. The solution provided by the data product is functional to support the development of higher-level solutions related to business problems, also known as business cases. The more versatile a data product is in supporting the development of solutions for multiple business cases, the greater its value.

To identify and then implement a valuable data product, it is first necessary to understand which business cases are relevant to the organization. To this end, it is essential to start with the analysis of the context in which the organization operates (business domain) and how it is structured internally to operate effectively within that context (business subdomains). In this section, we will show how to use **Domain-Driven Design** (**DDD**) to structure the problem domain, classify business cases of interest to the organization, and lay the foundation for solution modeling.

Introducing DDD

DDD is a software development methodology that focuses on creating a shared understanding of the problem domain within a development team and using that understanding to design and implement software systems. It was introduced by Eric Evans in his book *Domain-Driven Design – Tackling Complexity in the Heart of Software*. The overall goal of DDD is to create a model of the **problem space** that is both rich and expressive, allowing for a more effective and accurate translation of business requirements into software. By fostering collaboration between domain experts and developers and focusing on a shared understanding of a problem, DDD aims to reduce the risk of miscommunication and create more successful software systems.

Within an organization, the problem space aligns with the **business domain**. Complex organizations often span multiple business domains. Take Amazon, for instance, which operates across various domains, with retail and cloud services being its primary ones. Each domain can be further broken down into **subdomains**, allowing for the identification of smaller, more manageable portions of the problem space. These subdomains constitute cohesive and logically structured units, encompassing concepts, rules, processes, and managed business cases. While each subdomain addresses distinct business challenges, their solutions must be composable to support the organization's overarching goals. More complex subdomains can be further divided into nested subdomains, as shown in the following diagram (*Figure 4.1*).

Figure 4.1 – A retail domain example

The process of breaking down subdomains into finer-grained nested subdomains can be iterated until you have identified portions of the problem space with the right level of complexity for effective modeling.

Modeling the problem space (domain) or one of its subsets (subdomain) means creating an abstraction functional to solve the problems it contains. A robust model should contain precisely the information needed to address the problems it was designed for. In traditional software development methodologies, the domain model is often not explicitly expressed using a common language understandable to all stakeholders involved in a solution's design (*Figure 4.2*).

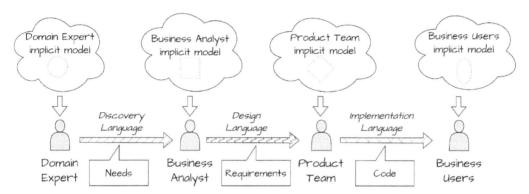

Figure 4.2 – A traditional approach to solution design

In the design process, the business expert typically translates their mental model of the domain into a series of needs. Using these needs as input, the analyst then constructs their own mental model of the domain, which they use to formalize a set of requirements. Subsequently, the development team utilizes these requirements to build their own mental model of the domain and subsequently translate it into code.

The concepts of needs, requirements, and code are all shaped by the individual mental models that each participant constructs and communicates to other actors involved in the design process, using their own specialized language. Within DDD, the focal point of the design endeavor is the creation of a shared model of a domain, articulated through a common language (*Figure 4.3*). This language is commonly referred to as **ubiquitous language** because it must be used pervasively by all involved actors, not only during design meetings but also in all generated artifacts (e.g., requirements, code, and documentation).

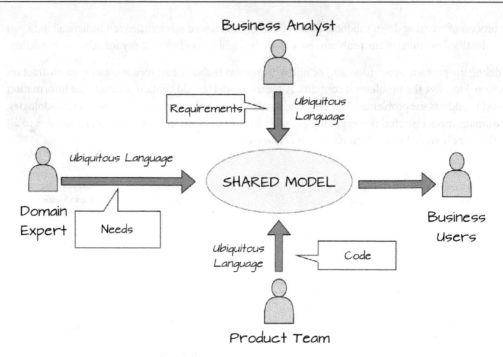

Figure 4.3 – A domain-driven methodology for solution design

Ubiquitous language is derived from business language, the language used by domain experts, eliminating all potential semantic ambiguities. Each term corresponds to a single concept, and vice versa, each concept is described by a single term. Clearly, it is practically impossible to define a single ubiquitous language to model a complex business domain. For example, consider how many meanings the term "customer" can have within an organization. In this case, giving a single definition to the term "customer" that contains only the elements common to all possible definitions would cause the ubiquitous language to lose the expressive power necessary to build useful models to solve real problems. Conversely, giving the term "customer" a definition that includes all possible definitions unambiguously would make the language too complex to be used effectively in modeling specific problems. Finally, giving up the use of the term "customer" and introducing new specific terms for each of its possible meanings would create a language that is functional to build a model but far from the language commonly spoken by a business. This would end up reproducing many of the typical problems of traditional design approaches.

In DDD, this semantic complexity is managed through the use of bounded contexts.

A **bounded context** is a well-defined conceptual and linguistic space in which terms, concepts, and rules have a specific, unambiguous, and consistent meaning. In other words, a bounded context defines the boundaries used to isolate a portion of a domain within which a specific ubiquitous language can be applied. In DDD, there is, therefore, not a single ubiquitous language; instead, there are as many ubiquitous languages as there are bounded contexts defined within the domain.

The term "customer" must have a single meaning within a bounded context, but it can have different meanings in different bounded contexts. The bounded context and its ubiquitous language determine the specific meaning to be attributed to the term "customer."

Domain modeling takes place within bounded contexts. There is no single language common to an entire domain, and therefore, there cannot be a single model. There are as many models related to specific subproblems as there are bounded contexts. If subdomains are a way to structure a problem space, bounded contexts are a way to structure a **solution space**.

Subdomains depend on the business domain in which an organization operates and the problems that characterize it. They are not defined. They are discovered through domain analysis. Conversely, bounded contexts are defined in a way that is functional to support producing solutions to the problems that characterize the business domain.

Ideally, subdomains and bounded contexts coincide. This occurs when there is a convergence between problems, models, and solutions. However, it is not always possible to have this perfect convergence between the structure of the problem space and the solution space. Let's see why and what the implications are in the next section.

Connecting problem and solution spaces

Subdomains determine the structure of a business domain that is functional for identifying and classifying problems of interest, also known as **business cases**. Conversely, bounded contexts determine the structure of the business domain that is functional for its modeling and the construction of solutions to identified problems. Subdomains are a structuring of the problem space, while bounded contexts are a structuring of the solution space.

Ideally, the structure of the problem space and the solution space tend to converge until they coincide. In other words, for each subdomain, there is only one bounded context, and vice versa. However, in reality, it may be necessary to generate different structures in which the relationships between subdomains and bounded contexts are not one-to-one, as shown in the following diagram (*Figure 4.4*).

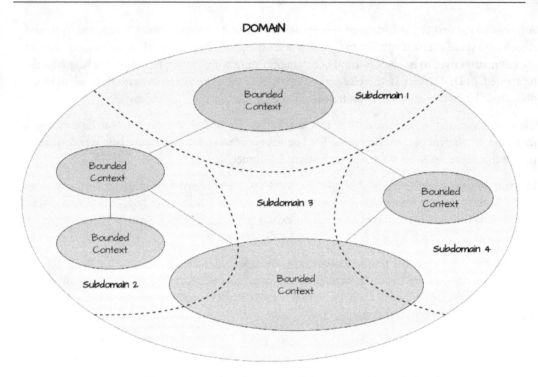

Figure 4.4 – Different relationships between subdomains and bounded contexts

The scenario where a bounded context contains multiple subdomains is typical of simple organizations where a single ubiquitous language can be used to model problems related to different subdomains. In this scenario, there is a risk associated with the growth of an organization. Growth, in fact, generates an increase in complexity within the subdomains. The language that was once common necessarily begins to articulate itself in divergent dialects within the various subdomains. However, by having a single bounded context, we try to solve these dialects within a single ubiquitous language and, consequently, with shared models and application solutions. These solutions inevitably become monoliths. As the complexity of the organization grows, the bounded context should, therefore, be divided into smaller bounded contexts, specific to the individual subdomains. Since the growth of complexity occurs slowly, it is not always easy to understand when the right time is to make this division. It is, therefore, easy to find ourselves having to intervene when it is too late, and the solutions developed within the common bounded context are now monoliths that are difficult to decompose into autonomous modules.

The scenario where a subdomain can contain multiple bounded contexts is instead typical of complex organizations where, even within the same subdomain, different languages can be used. In this scenario, there is a risk associated with the difficulty of associating use cases with application solutions. The organizational strategy is articulated in subproblems to be solved, the business cases, which are allocated to the various subdomains according to the structure of a business domain. Having multiple bounded contexts within the same subdomain can make it difficult to model the problems defined at the subdomain level. In general, the risk is that of having bounded contexts and, consequently, application solutions that are too interdependent. Each use case requires intervention on the models and applications defined in different bounded contexts. The result is poor modularization. The modules are present, but their coupling makes the solution as a whole a distributed monolith.

Whenever possible, it is always advisable to make the structure of the bounded contexts coincide with that of the subdomains, in order to align use cases, domain experts, and the team that will develop the solution (*Figure 4.5*).

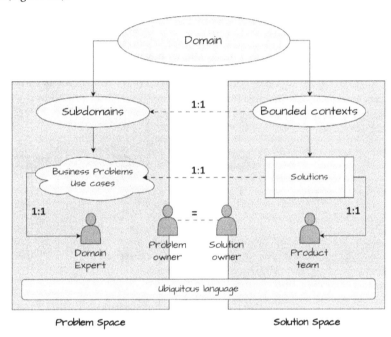

Figure 4.5 – Aligning bounded contexts and subdomains

This is the configuration that allows for decoupling between various solutions, separation of responsibilities among teams, and clear ownership. In other words, this is the configuration that promotes the best modularization of socio-technical architecture.

The structuring of the problem space into subdomains and the solution space into bounded contexts is the part of DDD called **strategic design**. In the next section, we will see how to structure the problem space by identifying its main subdomains.

Identifying subdomains

To identify subdomains and bounded contexts, it is generally useful to start by analyzing an organization's **business architecture**. **The Open Group Architecture Framework** (**TOGAF**) defines business architecture as follows: "*A representation of holistic, multi-dimensional business views of capabilities, end-to-end value delivery, information, and organizational structure; and the relationships among these business views and strategies, products, policies, initiatives, and stakeholders.*"

Business architecture is, therefore, a part of the entire socio-technical architecture of an organization. Specifically, it is the part that describes how the organization is structured to create, deliver, and capture value for its stakeholders. It includes various aspects of a company, such as its products or services, target market, revenue streams, cost structure, and overall strategy.

The **business model** is the starting point for defining business architecture. It describes at a strategic level what the organization intends to do to produce value for its stakeholders and how it intends to do it. Generally, the business model can be represented by means of a **business model canvas**.

In the rest of the chapter, we will use a hypothetical retail company called LuX as an example. LuX sells clothes and accessories from the most important luxury brands in its flagship stores. The LuX business model is described in the following business model canvas (*Figure 4.6*).

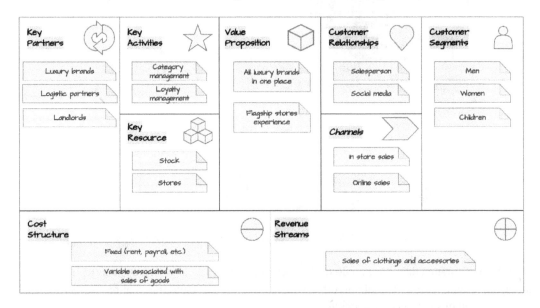

Figure 4.6 – The LuX business model canvas

Starting from the business model, it is possible to go down to a level of greater operational detail by identifying the macro activities or **business functions** involved in the value-generation process. This process is generally called the **value chain**. Within the value chain, each function contributes to creating a part of the value present in the products or services that an organization wants to offer to the market, in accordance with its business model. Typically, the functions that appear in a value chain can be divided into **primary functions** and **support functions** (*Figure 4.7*).

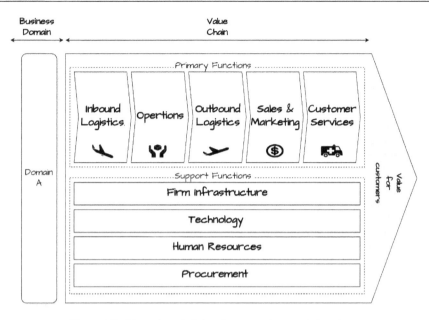

Figure 4.7 – The value chain's primary and support functions

Primary functions directly contribute to the production of the value that an organization wants to offer to the market in the form of products or services. Primary functions generally include **inbound logistics**, **operations**, **outbound logistics**, **sales and marketing**, and **customer services**.

Conversely, support functions indirectly contribute to the production of the value that an organization wants to offer to the market, in the form of products or services. Support functions generally include **firm infrastructure**, **technology**, **human resources**, and **procurement**.

Depending on the business model, some of these functions can be divided into more specific functions, merged, are not present, or otherwise classified as primary or support functions. For example, in the following diagram (*Figure 4.8*), it is shown in the LuX value chain how the procurement function is considered not a support function but a core function for the business. It is also merged with the inbound logistics function into a single functional unit. The marketing and sales function is instead divided into two autonomous functions. Finally, the outbound logistics function is not present.

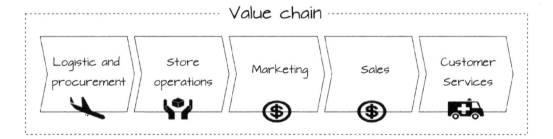

Figure 4.8 – The LuX value chain

The concept of the value chain was formalized by Michael Porter in his book *Competitive Advantage*, and it aims to explain how an organization produces economic value by differentiating itself in the market. It has a higher level of detail than the business model but still not enough to understand what the organization concretely does at an operational level to produce valuable outcomes for its stakeholders. To this end, it is necessary to go down one more level in the analysis and identify business capabilities and value streams.

A **business capability** is an ability that an organization has to achieve a specific goal or outcome. The functions that contribute to the value chain can be used to logically group business capabilities. In fact, each function must have a specific set of business capabilities to carry out the activities necessary to contribute to the value chain.

The following diagram shows the business capabilities of LuX and how they are distributed across the functions that make up its value chain (*Figure 4.9*).

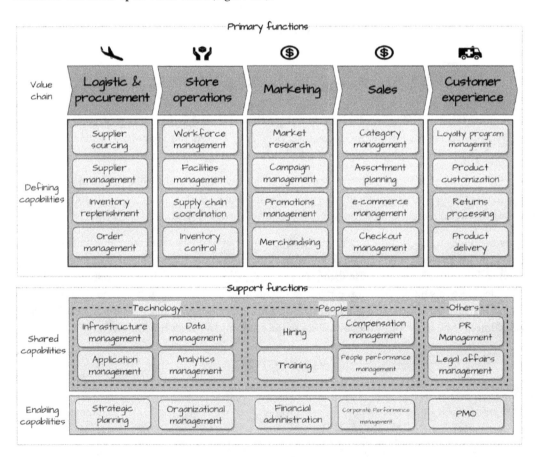

Figure 4.9 – LuX business capabilities

Each business capability can be decomposed hierarchically and iteratively into sub-capabilities of finer granularity. For example, in the previous chapter, we saw how the data management capability in a data product-centric architecture can be decomposed into the following four sub-capabilities – data product development, governance policy-making, XOps platform engineering, and data transformation enabling.

The value chain describes what an organization does at a high level to create competitive products and services in the market, in line with its business model. Business capabilities describe what the organization needs to be able to do to realize its value chain. **Value streams** describe how the organization uses its capabilities to perform end-to-end sequences of activities that contribute to the implementation of the value chain, producing tangible results for internal or external stakeholders.

In modeling terms, these value-adding activities are represented by value stream stages, each of which creates and adds incremental stakeholder value from one stage to the next. The following diagram shows the value stream that defines how LuX delivers a purchased product to the customer (*Figure 4.10*).

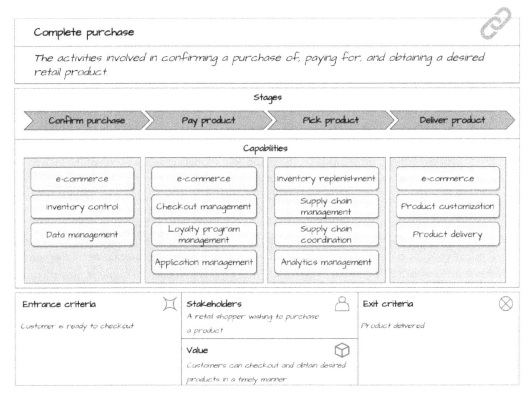

Figure 4.10 – The LuX complete purchase value stream

A business model, a value chain, a value stream, and business capabilities are elements of business architecture that are useful for describing what an organization does and why, at various levels of detail. However, they do not go into the details of how it happens and who does it. They are, therefore, useful tools for analyzing a problem space.

The hierarchy of functions, value streams, and business capabilities, once defined, can be used to map the domains and subdomains used in the DDD methodology. In particular, the finer-grained business capabilities can be used to define the finer-grained subdomains. In the following section, we will see how to structure a solution space by identifying its main bounded contexts.

Identifying bounded contexts

Organizational structure determines how tasks and responsibilities are distributed within an organization. The **operational units** defined at the organizational structure level implement business capabilities using the resources they have available (people, technologies, budget). The implementation of a business capability is called a **business capability instance**.

Generally, the organizational structure is derived from the functional decomposition carried out at the value chain level. Each operational unit is associated with a business function and has the task of implementing a specific set of capabilities related to it. However, the mapping is almost never one-to-one.

As the complexity of the organization grows, it is often necessary to modify the structure of the operational units to facilitate their management. Typically, this is done by specializing in one or more operational units by product type, sales channel, customer segment, and/or geography. For example, in a fashion company, the sales and marketing function could be decomposed at the organizational level into different operational units based on the brand and/or market segment, while the logistics function could be decomposed into different operational units based on geography. The aim is to reduce management complexity and make operations more aligned with the specificities of different markets.

This misalignment between the structure of the business functions of the value chain and the operational unit of the organizational model means that, in general, the same capability can have multiple instances at the operational level. It is also possible to have multiple instances of the same capability when an organization wants to distribute a specific activity across multiple operational units. For example, if the organization wants to manage strategic planning not in a top-down mode but in a distributed or federated mode, some of the capabilities associated with it will need to be instantiated on all the operational units involved. The same applies to the data management function. If, for example, the organization wants to distribute its management across all the main operational units while maintaining a federated approach, it needs to define a central data management function that instantiates the platform engineering, enabling governance capabilities, while it needs to create an instance of the data product management capability for each other operational function involved.

The hierarchy of operational units within the organizational model, and the related instances of business capabilities, are elements of the business architecture that are useful for describing at various levels of detail how an organization actually operates. They are, therefore, useful tools to analyze the solution space.

Once defined, the hierarchy of operational units and the instances of business capabilities associated with them can be used to map the bounded contexts used in the DDD methodology. In particular, bounded contexts can coincide with the different instances of business capabilities identified.

Defining bounded contexts based on business capability instances ensures alignment with the organizational structure and the right granularity to proceed with solution analysis. Alignment with the organizational structure is essential to easily identify who to involve in the analysis and with what responsibilities. In fact, it is not necessary to create roles and responsibilities orthogonal to the current organizational structure. The granularity of the business capability instances delimits an optimal perimeter to define a ubiquitous language, functional to the construction of a solution as indicated by the DDD. Finally, if the domains and subdomains are defined using the business capabilities and the bounded contexts using the business capability instances, the problem space and the solution space tend to coincide, as desired. The relationship between the subdomain and bounded context is, in most cases, one-to-one, and only in some cases is it one-to-many. For these reasons, it is better to think carefully before following other approaches to structuring the problem space and the solution space.

For example, when it comes to data management, it may make sense to define bounded contexts based on core business entities, such as the customer, the product, and the order. In practice, however, this decomposition orthogonal to the existing organizational structure complicates both the definition and management of solutions. It is more difficult to define solutions because it is not possible to rely on a single ubiquitous language common to an entire organization to define such entities. Similarly, it is more difficult to find a single business expert with all the knowledge necessary to model these entities. Finally, it is more difficult to maintain these solutions over time because it is necessary to define roles and responsibilities orthogonal to the current organizational structure, and then manage all the complexities related to the management of divergent objectives between the various work groups in a matrix organizational topology.

So far, we have seen how to use business capabilities and their instances to map the structure of the subdomains that constitute the problem space and the structure of the bounded contexts that constitute the solution space, respectively. In the next section, we will see how to identify business capabilities.

Mapping business capabilities

There are two main approaches to identifying the core business capabilities of an organization – *top-down* and *bottom-up*.

The top-down approach starts with the analysis of the business model and then moves on to the value chain and value stream analysis to define business capabilities. Once the business capabilities are identified, they are mapped to their corresponding instances at the organizational structure level. The top-down approach requires the active participation of top management.

The bottom-up approach starts with the organizational structure and then analyses the activities carried out by the different organizational units, arriving at the instances of business capabilities. Once the capability instances are defined, an organization's business capabilities are derived from them. The bottom-up approach requires the active participation of middle management. However, the participation of top management is still useful in the phases where the capability instances are grouped into generic business capabilities.

Both approaches to defining business capabilities can be carried out in a more or less iterative mode. At one extreme, you can try to build a complete map of all capabilities from the start (big-bang mode), or at the other extreme, capabilities can be identified progressively as needed (by opportunity mode).

Usually, a combination of approaches and modes according to a diagonal strategy is the best option. It starts by defining a set of core capabilities with a top-down approach and big-bang mode. Then, it proceeds to add capabilities of greater detail or lesser strategic relevance with a bottom-up approach, in by-opportunity mode (*Figure 4.11*).

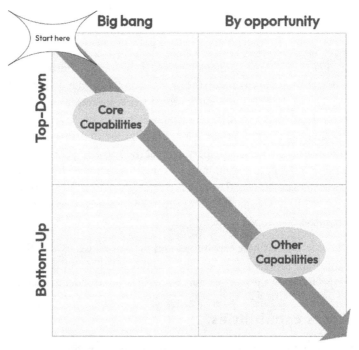

Figure 4.11 – Capability mapping approaches

In many organizations, the business architecture has already been defined in some or all its parts. In these cases, it is possible to reuse the artifacts produced, extending or updating them where necessary.

Discovering data products with event storming

Data products should not be developed with a stock-to-order logic. In other words, they should not be developed in the absence of clear business cases.

At the organizational level, business cases are defined based on the strategic plan. Some business cases require access to an organization's data assets. If there are no existing data products that make the data assets necessary for the business cases available and easily usable, then they must be implemented, or the existing ones must be extended.

In this section, we will see how to use the event storming technique to define the relevant business cases to implement a business strategy. We will then show how, also through event storming, it is possible to identify the data products to be developed to support defined business cases.

Understanding a business strategy

Data products are identified based on the relevant business cases. A business case is relevant if it supports an organization in achieving its business strategy. To define the business cases first and then identify the data products, it is, therefore, necessary to have a clear understanding of the business strategy and how it is built through the strategic planning process.

The strategic planning process begins with the leadership team (System 5) defining the vision that will guide the strategy. Subsequently, top management articulates the strategic vision into a series of goals that an organization must achieve to implement the vision (System 4). Middle management then defines the initiatives necessary to achieve the objectives defined by top management (System 3). Finally, the operating units identify the business cases useful to support the initiatives.

The strategic vision is defined based on the analysis of the context in which an organization operates and its current competitive positioning. This analysis identifies the organization's strengths and weaknesses. Different possible evolutionary scenarios of the external context are then analyzed. This analysis identifies potential opportunities and risks to which the organization is subject in the future. Thus, the information collected, together with the organization's purpose and principles, allows the leadership team to synthesize within the strategic vision the direction that the organization must follow in the future to compete in the market. Examples of tools to support this phase of strategic planning are market analysis, scenario planning, SWOT analysis, Wardley map, and Porter's five forces model.

Goals are defined by analyzing how an organization currently operates in order to identify what needs to be done to implement a vision. This analysis is generally conducted at the value chain or value stream level to identify what the organization currently lacks to implement the strategic vision. The goals identify the main gaps between the outcomes generated and those desired. They also indicate the expectations for improvement for each gap. Gap analysis is the common tool used to support this phase of strategic planning.

Initiatives are defined based on the goals, starting from the deficient elements at the value chain or value stream level. Specifically, it is analyzed how the organization must create, modify or reconfigure its capabilities to fill the highlighted gaps in order to achieve the established goals. Capability-based planning is the common tool used to support this phase of strategic planning.

Business cases are finally defined within the initiatives, based on the specific contribution that each of them is expected to make to manage the gaps that need to be filled, in order to implement the business strategy. A business case is a project or product that is useful for achieving the objectives of the initiative it is part of. It is a way to divide the business problem that the initiative is trying to solve into smaller problems that are easier to analyze and develop.

The process that leads from the definition of the business strategy, carried out at the management plane level (System 5–3), to its implementation at the operational plane level (System 1), and vice versa, is an incremental and iterative process that involves the entire organization. It is also a recursive process because once the business strategy for the entire organization has been defined, each operating function can in turn define its own strategy. The strategy of an operating unit must obviously be aligned with that of the entire organization.

The data management function can be involved in the strategic planning process at different levels. In organizations where data plays a key role in the business model and the production of competitive advantage, the data management function is represented at the leadership team or top management level. It is, therefore, involved in the strategic planning process, starting from the definition of the vision and the related goals and then contributing to all subsequent phases. At the opposite extreme, in organizations where data has a supporting role, the data management function is only represented at the operational middle management level. It is, therefore, not involved in the definition of the strategy but only in its implementation. In other words, it is involved in the strategic planning process only once the business cases have been identified to support its implementation.

Since data assumes an increasingly central role in every organization, the data management function is also gradually moving from a role of operational support to implement a strategy to a more active role in its definition. In most organizations today, the data management function is responsible for, or at least involved in, the definition of business cases based on the goals and key initiatives identified by top management.

The main phases of strategic planning are the same, whether the object of planning is the overall strategy of an organization or a local strategy of one of its operating functions. The incremental and iterative planning process is governed by the operating model that the organization or operating unit has chosen to adopt. Some organizations may adopt more formal and structured operating models, while others may choose more informal and agile ones, depending on the context. In *Part 3* of the book, we will see different operating models that the data management function can adopt to govern the process of defining and implementing a data strategy. In the next section, we will see how to translate the business strategy of an organization into an action plan (i.e., how to identify, starting from goals and initiatives, the priority business cases to be implemented at the operational level).

Turning business strategy into actionable business cases

Once top management has identified the priority initiatives to invest in, it is the task of middle management to define the business cases within each initiative to allocate the available budget. The identified business cases must clearly justify how and to what extent their solution would contribute to achieving the objectives of the initiative, and what the estimated costs to define and implement such a solution are. This information allows top management to prioritize the identified business cases and allocate the budgets associated with the individual initiatives among the business cases that compose them.

The initiatives aim to create, improve, or reconfigure a value stream or a specific business capability instance. To identify the business cases associated with an initiative, it is, therefore, necessary to first understand the stream or capability with which it is associated (i.e., to analyze the business processes that regulate its operation). Once it is understood how the value stream or capability is currently implemented at the operational level, it is necessary to identify opportunities for improvement. Among the possible opportunities for improvement, it is necessary to focus specifically on those that can potentially contribute most to achieving the objectives for which the initiative was defined. These are the ones that will constitute the business cases to be prioritized and financed. The analysis of the processes related to the problem at hand and the identification of the business cases of interest is conducted by the middle management responsible for the capabilities involved. In this phase, the objective is not to find solutions but simply smaller sub-problems that are worth investing in and working on. These analysis activities are generally conducted at the beginning of each annual budget cycle and then reviewed on a quarterly basis.

Before delving into the techniques that can be used to conduct this analysis, let's look at the case of LuX, the retail company we are using as an example. LuX sells clothes and accessories of the leading luxury brands. The main sales channel is a network of flagship stores located in the city centers of the countries in which LuX operates. During the COVID-19 pandemic, LuX also developed the e-commerce channel as an extension of the capabilities it already had to manage in-store sales. No dedicated capability instances were created to manage, for example, checkout or shipping on the e-commerce channel, but those already existing were reused. This allowed the company to quickly cope with the crisis situation without blocking the flow of sales and use underutilized resources, due to the drastic reduction in in-store sales.

After the pandemic crisis, LuX started working on its new strategic plan. The leadership team, based on a study of the competition and evaluating different scenarios of possible future developments of the retail market, became strongly convinced of the need to continue and even increase investments in the e-commerce channel. It thus summarized this conviction in the following vision statement to be used as a north star for the construction of a three-year strategic plan: "*We aim to become the best-in-class omnichannel luxury retail by providing a unique, personalized, and consistent customer experience across all channels.*"

LuX's top management then translated the vision into a series of goals. Many of the goals identified are linked to the leadership team's desire to evolve the e-commerce channel, while maintaining a unified customer experience with that traditionally offered in-store. As the strategic plan is defined over a three-year period, the goals have also been broken down by year. For the first year, one of the most important goals defined by top management is to improve the e-commerce shopping experience, which was rated as not excellent by customers in the company's periodic surveys. In particular, two initiatives have been identified as priorities for the e-commerce channel – improving the product search process within the catalog and improving the purchase completion process.

Both value streams have gaps compared to the competition and customer expectations. The expected improvement has been quantified by means of some success metrics. For example, for the improvement of the purchase completion process, the metrics used are the lead time, the number of complaints, and the judgment given by loyal customers in periodic surveys.

The heads of the functional areas whose capabilities are involved in the two value streams identified as deficient have planned analysis meetings, where they will determine the business cases to propose to top management to achieve the desired goals during the year.

In the next section, we will use the initiative related to the improvement of the "Complete purchase" value stream to see how the business processes that regulate its operational functioning can be analyzed by middle management, identifying business cases of interest.

Analyzing processes with event storming

A **business process** describes how a specific activity is performed at the operational level. In general, instantiating a business capability means defining the business processes that determine how the activity associated with the capability will be performed at the operational level and what resources it needs to be performed successfully (*Figure 4.12*).

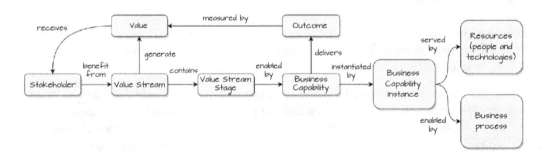

Figure 4.12 – Business architecture elements and relationships

A value stream, in turn, uses various capabilities grouped into different stages to produce the desired value for the reference stakeholders. At the operational level, therefore, a value stream is defined by combining the business processes that regulate the operation of the individual instances of business capabilities used by it. For this reason, whether you are analyzing an initiative associated with a single capability or an initiative associated with an entire value stream, it is always necessary to start from an analysis of the existing business processes in which you will need to intervene.

There are various techniques for analyzing business processes, including value stream mapping, domain storytelling, **Business Process Management** (**BPM**), and event storming. In this section, we will describe the latter because it is not only a lightweight but also a very flexible methodology that can be used for various purposes within the process of analysis, identification of opportunities, and design of solutions. Other methodologies such as BPM are instead more suitable for representing processes formally and managing their life cycle. Practices of this type can be used in a complementary way to the analysis done with event storming to document the processes and lay the foundations for their operational management.

Event storming is a workshop for analyzing one or more business processes defined by Alberto Brandolini. The workshop can have several purposes. Here, we will show how to use event storming to identify business cases. In the next section, we will see how to extend the analysis to define the solutions and, in particular, to identify the data products needed to support the implementation of the identified business cases.

In addition to a facilitator, the workshop is attended by business experts involved in the process or processes being analyzed. For the identification of business cases, the managers of the capabilities involved in the processes are generally directly involved. The workshop can be conducted both in person and virtually. During the analysis, a board, Post-its of different colors, and markers are used. We will not go into detail here on the methods and best practices to use to conduct these workshops. Event storming is a very widespread practice in the DDD community, and there is no shortage of high-quality online material on the subject. Some resources to start with in this sense are available in the *Further reading* section at the end of the chapter. Here, we will see what the main phases in which a workshop is divided are, as well as how to get from the analysis of a process to the identification of opportunities and problems to be translated into business cases.

In the first phase of a workshop, the participants list the events that occur as a process is executed. Each event is represented on the board by a Post-it of the same color, usually orange. This phase is performed in brainstorming mode, starting from a pivotal event and then definitions of the others, both upstream and downstream. Once the first set of key events has been defined, the board is reordered, duplicate events are consolidated, and ambiguous events are redefined.

In the second phase, an action or command is associated with each event. Each command is represented on the board by a Post-it of the same color, usually blue. At this point, the board is a more or less linear sequence of commands, followed by events that trigger other commands, and so on (*Figure 4.13*).

Figure 4.13 – An overview of events and command flow

In the third phase, the actors involved (yellow Post-its), the systems that support their execution if present (pink Post-its), and the policies that enable their execution (purple Post-its) are associated with each command. Policies are particularly useful to explicate the rules that govern the flow of commands and events that constitute the analyzed process.

Phases two and three, like phase one, are also composed of a brainstorming part followed by a rationalization part on the board. The three phases can be iterated several times until a consensus is reached on the quality of the representation of the analyzed process.

At this point, a brainstorming phase begins, with the aim of identifying problems and potential opportunities. Each participant adds a red Post-it to the board near where the problem or opportunity occurs, with a brief description. The red Post-its are then analyzed together and rationalized where necessary. At this point, each participant votes on the problems or opportunities that they consider most interesting to improve a process in the direction indicated by the goals associated with the initiative.

The following diagram summarizes the types of elements represented by the various Post-its used during the analysis and their relationships (*Figure 4.14*).

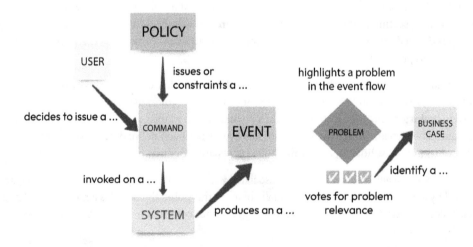

Figure 4.14 – Event storming components and relationships

The problems or opportunities that collect the most votes are finally used as the basis for the definition of the business cases. Each business case should be associated with a real business problem or opportunity. In particular, business cases associated with problems should be associated with the actual root problems (e.g., a lack of effective marketing campaigns), not symptomatic problems generated as a consequence of them (e.g., sales figures are down).

Each business case should also be **SMART** (**Specific, Measurable, Actionable, Realistic, and Time-bound**). To be specific, it must clearly define what is needed and how it is intended to proceed to build a solution. To be measurable, it must define how its solution will impact the goals of the initiative it is part of. To be actionable, it must clearly define the problem and lay the foundation for the definition of its solution. To be realistic, its solution must be implementable, considering the limitations of the resources available. Finally, to be time-bound, it must define the timing to address the problem and implement the solution.

All this information can be reported in a document with a standard structure, which is produced as an output of the analysis activity carried out.

In the case of LuX, the analysis of the processes that implement the "Complete purchase" value stream at the operational level, as shown in the following diagram (*Figure 4.15*), highlighted how the main problems are related to delays in shipping due to unexpected stock-outs of products.

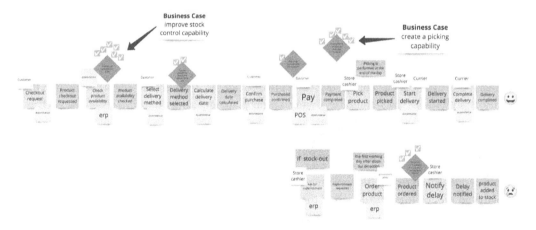

Figure 4.15 – An overview of the complete purchase process analysis

Since there is currently no system that manages an aggregate from which to derive a read model capable of exposing an updated inventory state, it is necessary to define a data product capable of implementing this logic by integrating different systems.

As a first step, the daily inventory update process needs to be analyzed to understand all the events that contribute to changes in the inventory state and the systems that generate them.

Using event storming, it is possible to first define the actions necessary to update the consolidated inventory. Each of these actions can then be associated with a read model containing the information needed to perform the update. The read models identified can finally be linked to the system capable of producing them.

The result of this analysis process is shown in the following diagram (*Figure 4.16*).

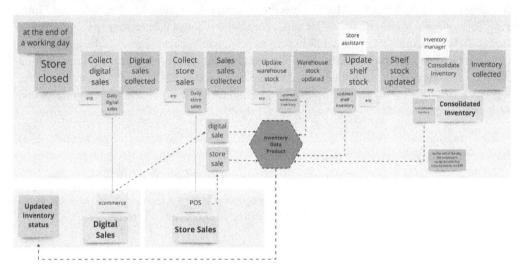

Figure 4.16 – An overview of the data product inventory

In summary, the inventory can change due to an in-store sale, an e-commerce sale, a variation in the quantities of products in the warehouse, or a variation in the quantities of products on the shelves. Sales in-store and on e-commerce platforms can be provided, respectively, by the e-commerce platform and **Point of Sale** (**POS**) systems. Changes in the quantities of products in the warehouse and on the shelves can be provided by the **Enterprise Resource Planning** (**ERP**) system.

The data product inventory to be implemented will, therefore, have four input ports, one for each identified read model, to capture each of these possible variations, and at least one output port to provide the updated inventory status. A fifth input port has been added during the analysis phase to capture the consolidated version of the inventory, calculated by the ERP system once a day. This is indeed the only version validated at the accounting level, and it is, therefore, necessary to realign with it as soon as it is calculated.

The data product inventory, thus defined, reads from multiple systems, consolidates the data, and produces an updated view of the inventory as required by the business case. However, the solution is not optimal because the data extracted from individual systems is processed before being redistributed – that is, there is a consumer-aligned product that reads directly from the sources. This could be a problem if the data read from one of these sources were to be needed by another product in the future. The latter would have to independently reimplement the data extraction part from the source. For example, the inventory product reads store sales data from POS systems but does not redistribute this data. If tomorrow the sales data was needed by another data product, the inventory product would not be helpful. For every potentially relevant data extracted from a source system, it is, therefore, preferable to define a dedicated source-aligned data product. In the case at hand, it was decided to implement, in addition to the consumer-aligned inventory product, four other consumer-aligned products, as shown in the following diagram (*Figure 4.17*).

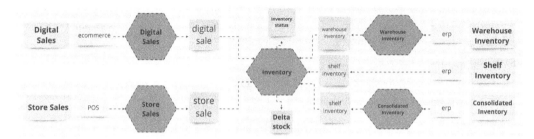

Figure 4.17 – Identified data products

The digital sales and store sales data products belong to the e-commerce and checkout bounded contexts, respectively. The warehouse inventory, consolidated inventory, and inventory products belong to the inventory control bounded context. For data related to the inventory of shelf products, it was decided not to implement a dedicated source-aligned data product because there are no foreseeable plans to reuse it in the future to support other business cases.

We have, thus, identified the data products needed to support the business case under consideration. In the next section, we will see how to describe them, place them, and manage them within a larger portfolio of data products.

Managing the data product portfolio

The data products defined based on the business cases become part of the broader portfolio of data products managed by an organization. The composition and evolution of this portfolio need to be managed in order to maximize value for the organization and minimize associated risks as much as possible. In this section, we will explore some key activities in the process of managing the data product portfolio.

Validating data product proposals

Before a new data product proposal is added to the portfolio, it must be validated. The validation process ensures that the data product complies with the principles and rules governing architecture development. Here are some examples of checks that can be performed at this stage:

- The data product must specify which domain it belongs to and who owns it.

- The data product must have a clear value proposition, linked to a specific business case if possible.

- A source-aligned data product must belong to the same domain as the source systems it reads from.

- A source-aligned data product should read data from a single source system. If this is not the case, it should be evaluated whether it is possible to divide the proposed data product into multiple products, one for each data source.

- A data product that reads from external sources should not also read from other data products. If this is not the case, it should be evaluated whether it is possible to divide the proposed data product into source-aligned products that only read from source systems and consumer-aligned data products that only read from other data products.

- A source-aligned product should read one aggregate from the source system to which it is associated. If this is not the case, it should be evaluated whether it is possible to divide it into multiple products, one for each aggregate.

- A source-aligned product should not apply business logic to the data it reads. It should only redistribute it, applying, at most, syntactic transformations. If this is not the case, it should be evaluated whether it is possible to divide it into a source-aligned product that returns the aggregate data and a consumer-aligned product that applies the desired business logic to the data.

- A source-aligned product should redistribute all relevant data related to the aggregate it reads from the source system. If this is not the case, it should be evaluated whether it is possible to extend it so that all relevant data of the aggregate is present in the output port.

- A consumer-aligned data product should not redistribute data read from another data product beyond what is necessary to ensure interoperability (e.g., keys and identifiers). If this is not the case, it should be evaluated whether the output ports can be modified to reduce the redistribution of data owned by another upstream product.

- A data product should not redistribute different views of the same data through different ports (e.g., one port for product stock for men and one for women). If this is not the case, it should be evaluated whether it is possible to define the ports in such a way that they can filter and profile the data according to the request or the requester (e.g., a single port for all stock that can accept a filter on gender).

The list is neither normative nor exhaustive. It is up to the governance capability of the data management function to define which rules make sense to check at this stage. Some rules can be blocking. In other words, if they are not respected, the product cannot be added to the portfolio. Conversely, others indicate guidelines to be followed and can be evaluated on a case-by-case basis.

In some cases, it may be necessary to add a data product to a portfolio even if it does not comply with all the established rules, due to budget or time constraints. In these cases, the non-respected rules should be noted in the data product description so that the non-conformities can be fixed at the first opportunity.

In the next section, we will see how to describe a product in order to add it to the portfolio once it has been validated.

Describing data product with data product canvas

Once a data product has been validated, it can be added to the portfolio. To do this, it is necessary to provide a high-level synthetic description of it. A **data product canvas** is a one-page document that contains all the key information needed to describe a data product. Multiple data product canvas templates are available online. They can be adopted as is or adapted to your needs. The following diagram shows a possible example of a data product canvas that describes the inventory product (*Figure 4.18*).

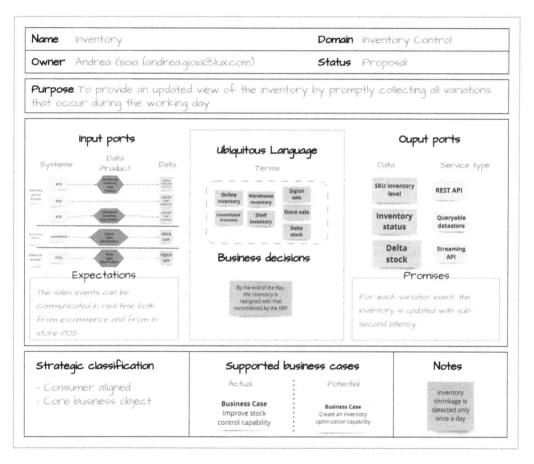

Figure 4.18 – An overview of a data product canvas

The data product canvas is a useful tool for communicating the value of a data product to stakeholders and managing the data product portfolio. It can also be used to track the progress of a data product throughout its life cycle.

The data product canvas typically includes the following information:

- **General information**:

  - **Data product name**: The name of the data product

  - **Data product owner and domain**: The person or team responsible for the data product and the domain it belongs to

  - **Data Product Purpose**: The purpose of the data product and the value it provides to the business.

  - **Data Product Status**: The development status of the data product (proposal, development, production, etc.)

- **Input**:

  - **What data does the data product read and from where (input ports)**: The data sources for the data product and the interfaces through which they are acquired

  - **What are the assumptions about the data consumed (expectations)**: The expectations about the data in terms of quality, format, completeness, and so on

- **Business logic**:

  - **What is the business logic with which the data is transformed and enriched (business decisions)**: The transformations and enrichments applied to the data to generate value

- **Output**:

  - **What data does the data product expose and through which interfaces**: The data made available by the data product and the interfaces through which it can be accessed

  - **What are the guarantees provided on the exposed data (promises)**: The guarantees provided in terms of the quality, reliability, security, and so on of the exposed data

- **Other details**:

  - **What is the product classification**: The classification of the data product based on its maturity level, criticality, and so on

  - **What are the business cases currently supported by the product, and what are the potential ones**: The business use cases currently supported by the data product and those that could be supported in the future

  - **Notes of various kinds**: Any additional notes on the known limitations of the data product, the validation criteria not being met, and so on

Optimizing a data product portfolio

The data product portfolio should be reviewed periodically to monitor its status and plan corrective actions where necessary.

Specifically, the following should be evaluated periodically:

- **Existence of products with similar purposes that can be merged**: If there are products that offer similar value and functionality, it may be possible to merge them into a single product, simplifying the portfolio and reducing maintenance costs.

- **Products that no longer support any business case and can therefore be decommissioned**: Products that are no longer used or that no longer provide value to the business can be decommissioned to free up resources and reduce costs.

- **Products for which the classification type needs to be changed and, consequently, the management methods**: A product that was born as a tactical solution could, for example, become strategic, depending on the number of new business cases it supports. In this case, the validation rules to which it must conform become more stringent. Exceptions that could be tolerated before must now be addressed.

- **Structurally similar products that may suggest the development of a platform component or service to facilitate the future development of other products of the same type**: By identifying commonalities between products, it may be possible to develop reusable components or services that can streamline the development process and reduce costs.

- **Products that do not meet the consumption guarantees provided (promises) and must, therefore, be quarantined**: Products that do not meet the agreed-upon quality, reliability, or security standards must be quarantined to prevent them from impacting other products or users.

- **Products that have undergone structural changes over time for which the documentation needs to be updated**: It is important to keep documentation up to date to ensure that stakeholders have accurate information about a product and its capabilities.

- **Imbalance in the type of products present in the portfolio that requires intervention**: For example, there may be too many products in a generic or support domain and too few in a strategic domain. This imbalance can be addressed by developing new products in the strategic domain or by decommissioning products in the generic or support domain.

- **Products that need to be repositioned in a new domain following organizational changes**: If an organizational structure changes, it may be necessary to reposition some products in a new domain to reflect the new structure.

- **Products for which a change of ownership needs to be managed**: If the ownership of a product changes, it is important to ensure that the new owner has the necessary knowledge and resources to manage the product effectively.

- **Low reuse of data products to support business cases different from the one for which they were originally developed**: The analysis should be intensified in this case to understand how to intervene, in order to increase the composability of data products.

By regularly reviewing the data product portfolio, organizations can ensure that it remains aligned with their business needs and that it delivers the expected value.

Summary

In this chapter, we have seen how to identify a data product, starting from a business case. We first saw how to use domain-driven design to model a problem space, in order to find the problems of interest and then the best solutions. The problem space for an organization coincides with its business domain. To simplify its analysis, it is necessary to decompose it into smaller and easier-to-model parts. Starting from the business architecture, we used business capabilities and their instances at the operational structure level to determine the boundaries of the various parts of the problem space and the solution space, respectively.

We then used event storming to analyze the business processes of interest to the business strategy, in order to determine the business cases useful for its implementation. Again, using event storming, we saw how to identify data products from business cases.

Finally, we saw how to manage the overall portfolio of defined data products, in order to maximize the value, they produce for an organization, while reducing risks and complying with the rules defined at the governance level.

In the next chapter, we will see how to move from identifying a data product to developing it.

Further reading

For more information on the topics covered in this chapter, check out the following resources:

- *Domain-Driven Design: Tackling Complexity in the Heart of Software* – Eric Evans (2003): `https://www.amazon.com/Domain-Driven-Design-Tackling-Complexity-Software/dp/0321125215`

- *Domain-Driven Design Starter Modelling Process* – DDD Crew: `https://github.com/ddd-crew/ddd-starter-modelling-process`

- *Business model generation* – A. Osterwalder, Y. Pigneurr (2010): `https://www.amazon.it/Business-Model-Generation-Visionaries-Challengers/dp/0470876417`

- *Competitive Advantage* – Michael E. Porter (2004): `https://www.amazon.it/Competitive-Advantage-Michael-Porter/dp/0743260872`

- *Value Streams* – TOGAF® Series Guide (2022): `https://pubs.opengroup.org/togaf-standard/business-architecture/value-streams.html`

- *Business Capabilities* – TOGAF® Series Guide (2022): `https://pubs.opengroup.org/togaf-standard/business-architecture/business-capabilities.html`

- *Business Capability Planning* – TOGAF® Series Guide (2023): `https://pubs.opengroup.org/togaf-standard/business-architecture/business-capability-planning.html`

- *Introducing EventStorming* – A.Brandolini: `https://www.eventstorming.com/book/`

- *EventStorming Glossary & Cheat sheet* – DDD Crew: `https://github.com/ddd-crew/eventstorming-glossary-cheat-sheet`

- *Data Product Flow* – P.Platter (2022): `https://www.agilelab.it/knowledge-base/how-to-identify-data-products-welcome-data-product-flow`

- *The art and science of data product portfolio management* – F.Haddad and M.Mayrhofer (2023): `https://aws.amazon.com/blogs/big-data/the-art-and-science-of-data-product-portfolio-management/`

Get This Book's PDF Version and Exclusive Extras

Scan the QR code (or go to `packtpub.com/unlock`). Search for this book by name, confirm the edition, and then follow the steps on the page.

Note: Keep your invoice handy. Purchases made directly from Packt don't require one.

5

Designing and Implementing Data Products

Building on the previous chapter's discussion on identifying data products aligned with business strategy, this chapter delves into the development phase.

We'll explore how a data product integrates within the organization's technological landscape, interacting with other data products and application components.

We'll then dissect the core components of a data product, outlining their development process and formal description through machine-readable descriptor files.

Finally, we'll embark on a journey through the data flow within a data product, examining components responsible for sourcing, processing, and serving data.

This chapter will cover the following main topics:

- Designing data products and their interactions
- Managing data product metadata
- Managing data product data

Designing data products and their interactions

To ensure a successful data product implementation, we must first grasp its interaction with both the broader global ecosystem and its specific local environment. We'll also need to identify the core components that define its internal structure. In this section, we'll begin by delving into the local context, then take a step back to examine the global ecosystem, before finally zooming in on the product itself and its internal components.

Understanding the data product local environment

A data product supports one or more business cases by making a business data asset accessible and easy to use. Each data asset is generated as a consequence of activities carried out by an instance of business capability within a value stream. Data assets are, therefore, associated with the instances of business capability that generate them, which, in turn, are used to define the bounded context. Each data product is defined and implemented within the bounded context of the data asset that it manages and makes available for use. Its modeling and implementation use the ubiquitous language of the bounded context to which it belongs.

Data generated by the activities carried out by a capability instance can be grouped together to form cohesive and consistent entities within the business domain. In **domain-driven design** (DDD), these entities are called **aggregates**. Operational applications manage the creation and transactional evolution of aggregates in order to provide support functionality for operational processes.

Source-aligned data products read aggregates from operational applications and make them reusable to support multiple business cases, both analytical and operational ones. In operational applications, data associated with the aggregate is a by-product; in source-aligned data products, it is the product itself. Unlike operational applications that expose functionalities that, as a side effect, can modify the aggregate, source-aligned data products directly expose the data related to the aggregate that they manage.

Consumer-aligned data products enrich the aggregate data exposed by source-aligned products. All data products, both source-aligned and consumer-aligned, expose the data they manage in read-only mode. That is, they do not expose functionalities such as operational applications to modify them. The data can be exposed for consumption within the subdomain or for consumption by other subdomains, as shown in the following diagram (*Figure 5.1*):

Subdomain

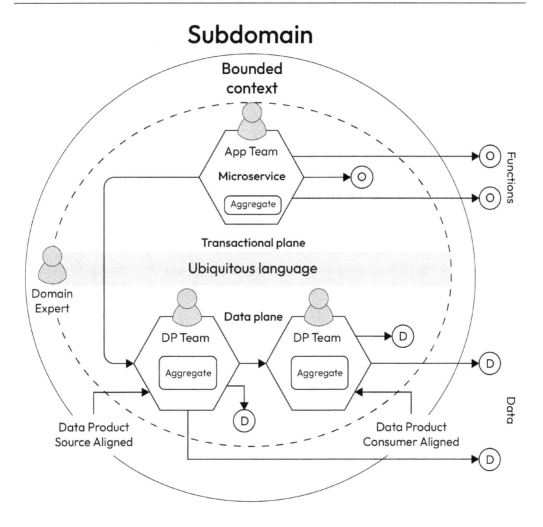

Figure 5.1 – Data product local context

A bounded context represents the local environment within which a data product is developed. Each data product has an owner and a team dedicated to its development. Generally, all products of a bounded context are developed by the same team, which can be internal to the operational function of which the bounded context is a part or belong to a centralized data management function depending on the structure that the organization has given itself. However, a domain expert must always be present to support the team in the development of the product.

Within the bounded context, the team that develops the data products collaborates with the team that develops the applications. Within the bounded context, both teams use the same ubiquitous language and interact with the same domain experts. In some cases, it is also possible that the two teams, application and data, coincide.

If applications are developed within the bounded context, following (for example) a microservices-based architecture, it is generally the case that each aggregate is managed by a specific application's module or microservice. Consequently, each microservice has a specific source-aligned data product associated with it. The functionalities exposed by the microservice allow the aggregate to be manipulated. The associated source-aligned data product reads the data of the aggregate and exposes it to support business cases internal and external to the subdomain.

To support the widest possible range of present and future business cases, a source-aligned data product should be able to update its data whenever the aggregate it is associated with is modified by the microservice or application that manages it at an operational level. This way, it is possible to support both real-time analytics and application integration business cases, not just batch analytics business cases. To this end, it is important that all significant events associated with an aggregate, identified (for example) during an event storming session, are exposed by the microservice or application that manages the aggregate and immediately acquired by the source-aligned data product.

An event describes a fact that occurred at a specific moment in time that led to a change in the state of the aggregate of interest. Events are immutable by nature. Once an event has taken place, it cannot be changed retroactively.

Of course, the modification made by an event to the state of an aggregate is not irreversible in itself. Subsequent events can further modify the state by canceling the modifications made by a previous event. For example, it is not possible to modify the fact that a customer placed an order at 2:00 P.M. today, which consequently modified the inventory state. However, it is possible to cancel the modification to the inventory state by generating a subsequent order cancellation event at 3:00 P.M.

Events and state are two sides of the same coin. Events cause changes to the state of an aggregate, while the state reflects the consequences of the events that have occurred. By knowing the state of an aggregate at a specific moment in time (snapshot) and all the events associated with it that have occurred since that time, it is possible to reconstruct the current state of the aggregate. The reverse is not true. Knowing the current state does not allow us to trace back the sequence of events that generated it. Events, therefore, have a greater information content because they not only allow us to build the current state but also allow us to track the sequence of changes that generated it.

To maximize the potential for reuse, a data product should, whenever possible, not only acquire data in real time as a sequence of events but also reshare the enriched data both as the overall state of the data asset and as the sequence of events that led to that state, immediately consumable by downstream data products.

Of course, managing data as events in order to guarantee consumption not only in batch mode, in addition to increasing the potential value of the data product, also increases its complexity. It is, therefore, always necessary to evaluate whether the potential return on value justifies the higher costs associated with the development and maintenance of the product.

We will discuss in more detail the methods for managing data as sequences of events within a data product in the following sections, dedicated specifically to sourcing, processing, and serving data.

Composable architectures

According to Gartner, a composable architecture is an architecture built on independent, interchangeable, and combinable systems and applications.

A **packaged business capability** (**PBC**) is the modularization unit of such architectures. For Gartner, a PBC's level of detail should directly correspond to a well-defined business capability within the organization. In this way, each PBC is clear and understandable even for non-technical users.

There are two flavors of PBCs: **application** (**A-PBC**) and **data** (**D-PBC**). An A-PBC provides functionalities that directly support a business capability, while a D-PBC offers data assets managed by that capability.

Following the approach proposed in this book of using business capabilities instances to define bounded contexts, the combination of all the data products within a single bounded context forms what Gartner calls a D-PBC. Likewise, the combination of all microservices within that same bounded context forms an A-PBC. The interface of these PBCs is given by the union of the APIs that the microservices or data products expose for use outside of their subdomain.

In order to reduce delivery delays due to unforeseen stockouts, LuX has decided to implement an *Inventory Data Product* within the *Inventory Control subdomain*. This product will expose the most up-to-date possible state of the inventory.

To implement this consumer-aligned data product, it is first necessary to implement several source-aligned data products capable of intercepting and exposing events generated by the operational systems that impact the inventory state. For example, each sales event modifies the inventory state by reducing the stock quantities of the products sold.

Since there are two sales channels, the store and e-commerce, and each manages sales using different operational systems, it is necessary to implement two separate source-aligned data products. The first, the *Digital Sales data product*, will be responsible for intercepting sales made on the digital channel. The second, the *Store Sales data product*, will be responsible for intercepting sales made on the physical channel.

The development of the *Digital Sales Data Product* is assigned to the in-house team that follows the development of the e-commerce platform. The ownership of the new data product is assigned to the e-commerce product owner, who will also act as the business expert during development.

The development of the *Store Sales Data Product* is assigned to a central team that reports to the data management function within LuX IT. The ownership of the new product is assigned to the head of the *checkout management capability* within the *sales function*. The same person will act as the domain expert and will be appropriately supported by the person responsible for the **point-of-sale** (**POS**) systems within central IT.

The following diagram (*Figure 5.2*) shows the *Digital Sales Data Product* within its local environment, the e-commerce bounded context:

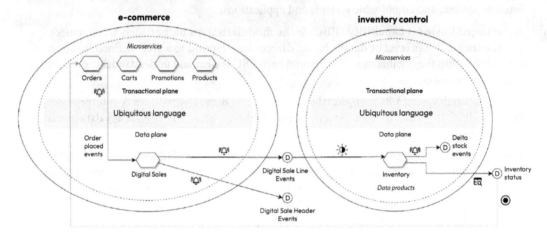

Figure 5.2 – Digital Sales Data Product's local environment

The e-commerce solution in LuX was implemented using a microservices architecture. The microservice that handles sales is the *orders microservice*. The new *Digital Sales data product* will need to acquire each order from the latter, process it appropriately, and expose it for consumption outside the *e-commerce subdomain*.

The acquisition, processing, and exposure of orders must be done in real time as much as possible to allow the downstream *Inventory Data Product* to expose an image of the inventory that is as up to date as possible.

The diagram maps how data is exchanged between operational systems, source-aligned data products, and consumer-aligned data products, both within a subdomain and between different subdomains.

Following the flow, we can see how the *Orders Microservice* exposes orders as a sequence of events, and the *Digital Sales Data Product* acquires them in real time as soon as they are exposed.

Once the events related to the issued orders have been acquired, the *Digital Sales Data Product* breaks down each order into a header and individual order lines and exposes these two pieces of information as events for downstream consumers.

The *Inventory Data Product* consumes the events related to the order lines to update the stock and expose its current state. The latter is read punctually by the *Orders Microservice* to decide whether a checkout request can be finalized based on the availability of the ordered products.

Since the *Inventory Data Product* also processes data in near real time, it is not difficult to expose individual stock variation events in addition to the inventory state. At the moment, there are no consumers for this data, but given the low effort involved, it makes sense to expose stock variations as well in order to increase the potential for reuse of the data product in the future and, consequently, its value.

The following diagram (*Figure 5.3*) shows a legend of the icons that can be used to map the ways in which data exchanged between applications and products is exposed and consumed:

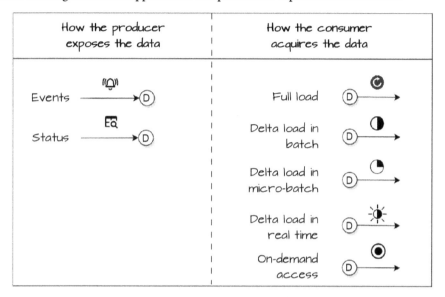

Figure 5.3 – Data exchange modalities

We have seen how a data product interacts with its local context or the bounded context within which it is developed. In the next section, we will zoom out to see how data products interact with the global ecosystem by operating outside of their bounded context.

Understanding the data product global ecosystem

In addition to interacting with other services and products within its bounded context, a data product may need to interact with external products or applications.

Applications and products external to the bounded context with which a product may need to interact can belong to other bounded contexts or be external systems; that is, systems that do not belong to any specific bounded context.

Examples of external systems are all **commercial off-the-shelf software** (**COST software**) used by the organization, monolithic legacy applications, and third-party services. Even when the services exposed by these external systems are used only within a specific bounded context, they are not part of it because they do not use its ubiquitous language.

In most cases, external systems offer functionalities used by multiple subdomains, as shown in the following diagram (*Figure 5.4*):

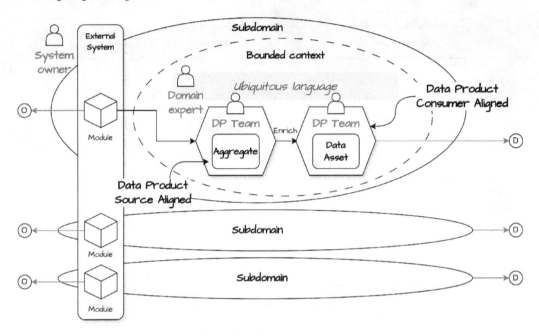

Figure 5.4 – External systems

In the case of LuX, the main external systems that need to be interfaced with in order to implement the products necessary to support the *real-time inventory business case* were identified during the analysis conducted using the event storming technique and represented by means of pink Post-its. Specifically, these are the **enterprise resource planning (ERP)** and the POS systems.

The ERP system is a legacy product developed in-house over time. It offers multiple functionalities, including consolidated stock calculation, warehouse management, and shelf-level merchandise tracking. Unlike e-commerce, the ERP system was not developed in a modular way, clearly distributing the different functionalities across the bounded contexts and aligning them with a specific ubiquitous language. Therefore, the ERP system, even though developed in-house, should be considered for all intents and purposes as an external system.

The POS software that manages the cash registers is a COST software installed in all stores. For the *real-time inventory business case*, it will be necessary to read the stock variations in the warehouse from the ERP system and at the end of the day realign with the consolidated inventory. It will also be necessary to read the individual sales made in-store from the POS system. For each of these interactions, a specific source-aligned data product will be developed, responsible for acquiring the data of interest from the external system and making it available for inventory update.

In addition to interacting with external systems, each data product may also need to interact with data products or applications present in other bounded contexts. This interaction can take place both to consume and to expose data from and to other bounded contexts.

For example, in the case of LuX, the *Inventory Data Product*, which belongs to the *Inventory Control Bounded Context*, reads the events useful for updating the inventory not only from the *Warehouse Inventory Data Product* and *Consolidated Inventory Data Product* belonging to its own bounded context but also from the *Store Sales Data Product* and *Digital Sales Data Product* respectively belonging to the *e-commerce bounded context* and *checkout bounded context*.

The *Inventory Data Product* also reads events related to changes in shelf-level products directly from the external ERP system. Finally, the *Inventory Data Product* exposes the updated inventory state to all other bounded contexts. In particular, the *e-commerce Bounded Context* uses this data to process orders, reducing disservices caused by unforeseen stockouts.

The following diagram (*Figure 5.5*) shows a map of the products needed to support the *real-time inventory business case* and their relationships:

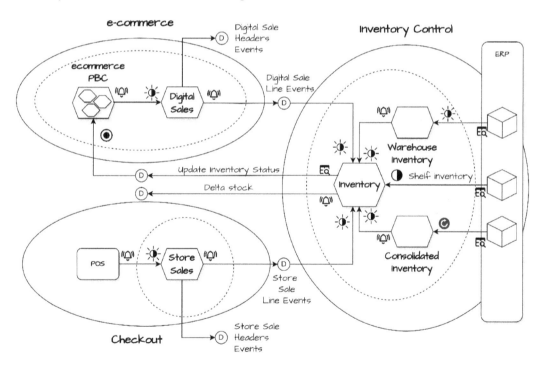

Figure 5.5 – Data product map

Mapping the relationships between data products and, consequently, between bounded contexts is not enough. It is also necessary to decide the logic according to which these relationships will be implemented. In fact, the integration mode determines how much two data products are coupled and, consequently, how much their respective bounded contexts are coupled. Although it is possible to decide the integration logic at the level of individual data products, it is generally more useful to do so at the level of bounded contexts. In fact, it is sufficient that there is a strong coupling between a data product of one bounded context and that of another to ensure that the two bounded contexts are strongly coupled, even if the coupling between the other data products is low or even non-existent. Furthermore, since two bounded contexts could be associated with different development teams, the interaction logics between bounded contexts determine not only how the products will be integrated but also how the respective teams will interact.

In DDD, the process of defining integration methods between bounded contexts is called context mapping. It is considered a part of strategic design. DDD codifies several standard patterns for integration between bounded contexts and how to represent them within a context map. These patterns describe multiple perspectives, such as service provisioning, model propagation, and governance aspects. The diversity of perspectives allows for a holistic view of the relationships between bounded contexts and their associated teams.

Within a context map, a relationship between two bounded contexts is represented by a straight line connecting them. If changes to the data products of one bounded context affect the other and vice versa, the two bounded contexts and their respective teams are mutually dependent. If, on the other hand, changes to the data products of one bounded context impact the other but not vice versa, the relationship is upstream-downstream. In this case, the line connecting the two bounded contexts is annotated with U near the upstream bounded context and D near the downstream bounded context.

If a relationship exists between two bounded contexts, it necessitates interaction between them at both the technological and organizational levels. The interaction patterns defined in DDD that are particularly relevant in architectures focused on managing data as a product are **Open Host Service (OHS)**, **Shared Kernel (SK)**, **Conformist (CF)**, **Anti-Corruption Layer (ACL)**, and **Published Language (PL)**.

The OHS pattern occurs when a bounded context exposes its data or functionality through services based on standard interfaces and protocols.

In the case of architectures based on managing data as products, it is desirable that this pattern always be implemented. The governance capability has the task of defining interfaces and protocols to ensure technical interoperability of data products. Development teams should implement their data products following these specifications. To indicate that this pattern is implemented, the line used to specify the relationship is annotated with an OHS string.

The SK pattern occurs when two bounded contexts share responsibility for the development of a component used by both. This could be an infrastructure resource, a utility service, or even an entire data product.

In the case of architectures based on managing data as products, it is desirable that this pattern never be implemented. It should never happen that multiple teams have the responsibility to develop the same data product. There can be resources or services shared between bounded contexts. In fact, to increase the scalability of the solution, it is desirable that there be shared resources and services to reduce the development and maintenance costs of data products. However, the responsibility for these shared resources and services should not lie with the product teams but with the platform capability and consequently with the team or teams responsible for it. To indicate that this pattern is implemented, the line used to specify the relationship is annotated with an *SK* string.

The CF pattern occurs when a downstream bounded context uses the data model adopted by the upstream bounded context without converting it into a model expressed through its own ubiquitous language.

In the case of architectures based on managing data as products, this is the standard pattern of interaction between products belonging to the same bounded context. However, it should be limited between products belonging to different bounded contexts, especially if they use divergent ubiquitous languages and are developed by different teams. To indicate that this pattern is implemented, the line used to specify the relationship is annotated with a *CF* string.

The ACL pattern occurs when a downstream bounded context develops a layer to isolate the data read from the upstream system and translate it into a model based on its own ubiquitous language.

In the case of architectures based on managing data as products, it is desirable that this pattern always be used when interacting with external systems. It can also be used in the case of interactions with products from other bounded contexts. When dealing with data products that expose strategic assets, prioritize the combined use of this pattern and the PL pattern (explained next) over the CF pattern. This ensures better scalability in the long run. To indicate that this pattern is implemented, the line used to specify the relationship is annotated with an *ACL* string.

The PL pattern occurs when two bounded contexts interact using a common language that is different from their respective ubiquitous languages. The upstream bounded context translates its model into the common language and exposes it through standard interfaces following the OHS pattern. The downstream bounded context reads the model described according to the common language and either conforms to it or translates it into a model described according to its own ubiquitous language using the ACL pattern.

In the case of architectures based on managing data as products, it is desirable that this pattern always be used to manage the integrations of strategic data assets whose relevance crosses the various business functions and capabilities (for example, customer, contract, product, and so on). In *Chapter 10, Distributed Data Modeling*, we will see how to define a common language for modeling strategic data assets through an enterprise ontology. To indicate that this pattern is implemented, the line used to specify the relationship is annotated with a *PL* string.

The following diagram shows a context map for the bounded contexts involved in the *real-time inventory business case* (*Figure 5.6*):

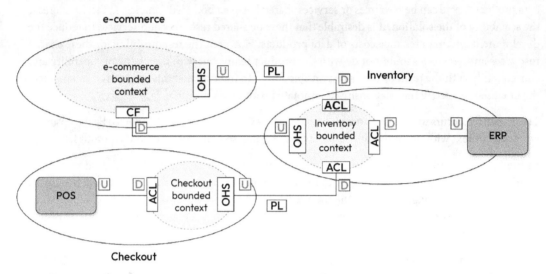

Figure 5.6 – Context map

All upstream bounded contexts expose data and functionality through the OHS pattern. Downstream bounded contexts access external data or functionality by applying the ACL pattern. The only exception is the *e-commerce bounded context*, which conforms to the *Inventory bounded context*'s model when reading inventory status. In this case, this is not a problem because the read is punctual and done for operational purposes. The data is not saved, and therefore the inventory model does not propagate within the *e-commerce bounded context*.

Sales, both digital and in-store, are core data assets that will be able to serve multiple business cases beyond the one under consideration. For this reason, it was chosen to expose them using a model based on a common language different from the one used in the *e-commerce bounded context* and *Checkout bounded context*.

Finally, the positioning of the external systems on the map indicates that the POS system is used only within the *Checkout domain*, while the ERP system also supports other domains besides the *Inventory domain*.

Having completed the analysis of the local context and the global ecosystem in which the data products we will develop are located; it is now time to delve deeper into the internal structure of a data product in the next section.

Understanding data product internals

Each data product has a unique name, belongs to a domain, and is managed by a product owner. Once developed, a data product can be deployed in one of the runtime environments present within the organization's application landscape (for example, dev, quality, pre-production, production). A data product can have multiple *data product instances*. Each instance is associated with a specific version of the data product. Generally, within a specific runtime environment, it does not make sense to have multiple instances of the same version of a data product. However, it may make sense to have instances associated with different versions; for example, to manage a blue-green deployment in which the old version coexists with the new one and consumers are gradually migrated from the old to the new one.

Each time a product is created or modified, a new version is released and an instance is created in one of the runtime environments. The runtime environment in which to instantiate a new version of a product, as well as the methods for subsequently instantiating it in other environments, are defined by the release policies that the organization has adopted. The same applies to the methods and timing of decommissioning old versions. The current version of a data product is the most recent version instantiated in a production environment.

A data product exposes a data asset for consumption. The content of the data asset is described by means of a dataset. Hereinafter, we will use the term "dataset" to refer to the conceptual model that describes a data asset at a semantic level. A dataset is instantiated at a physical level by means of one or more datastore. The level of denormalization of a dataset in one or more datastores depends on the physical modeling choices and the underlying storage system used to store the data asset data. A data product can also store the data of the same data asset in different storage systems to support different types of workloads and consumers. The same data can, therefore, be replicated in different data stores.

In addition to the datastores that physically manage the exposed dataset (**consumption datastores**), a data product can have other datastores used internally to store and process intermediate data (**inner datastores**) useful for producing the final datastores exposed for consumption. Inner datastores are not accessible to external consumers (information hiding).

Consumers should normally not directly access the consumption datastores of a data product. Access to data should always be provided through dedicated **data services** that expose interfaces defined according to common specifications (OHS). The output ports of a data product are implemented through data services that read data from the consumption datastores and expose it through appropriate views encoded in different formats. The data views exposed by a data service are generally called **data distributions**.

The application processes that transform the data acquired by a data product and transform it into the data exposed to consumers are called **data pipelines**. Data products acquire data of interest through services that implement their input ports. These services can acquire data in both push and pop modes. Once the data is acquired, it is normalized (ACL) and stored in inner datastores called **landing datastores**.

Data pipelines represent the business logic implemented by the data product to enrich the input data and make it ready for consumption. They should, therefore, be isolated from the acquisition and distribution methods. In other words, data pipelines have a clear responsibility: reading data from landing datastores and transforming it for storage in consumption datastores. Following the principles of hexagonal architectures, the task of acquiring data to be stored in the landing datastores and redistributing data stored in the consumption datastores should be delegated to appropriate adapters. Over time, adapters can be factored out, and their management and evolution can be delegated to the platform team.

The following diagram shows the main components that make up the internal structure of a data product (*Figure 5.7*):

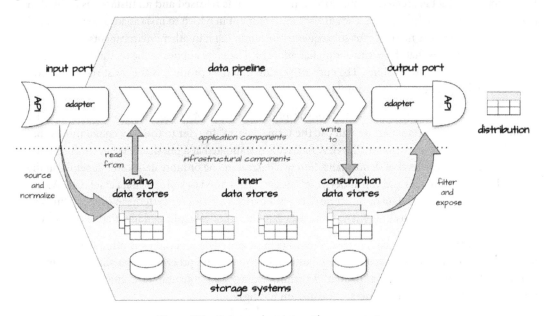

Figure 5.7 – Data product internal components

In the next section, we will see how to formally describe all the components of a data product using a descriptor document.

Managing data product metadata

A data product is not limited to managing and exposing data but must also manage and expose metadata that facilitates usability for consumers and operationalization by the platform. In *Chapter 2, Data Products*, we introduced the concept of data product descriptor document as a tool for collecting and sharing metadata related to all the components that make up a data product. Let's now see how to define this document and how to manage its life cycle.

Describing data products

A data product descriptor is a document that formally and in a machine-readable way describes all the components of a data product. It contains all relevant metadata about the product. It does not just describe the data it exposes and how it is exposed (interfaces) but also contains metadata about its internal components (applications and infrastructure components).

The metadata about the interfaces is publicly accessible and serves consumers to decide whether and how to use the data product. Generally, the part of the descriptor file that describes the interfaces is called a **data contract**. The metadata that describes the internal components is only visible to the development team and the self serve platform. This metadata is used to manage the product throughout its life cycle, from initial deployment to final decommissioning.

One of the first tasks of the governance capability when adopting a data-as-product approach is to define what a data product is and how to formally describe it using a descriptor document based on a shared specification. The presence of such a specification ensures, on the one hand, technical interoperability between data products and, on the other hand, lays the foundation for the development of platform services that facilitate their operational management.

A good specification for describing a data product should be the following:

- **Expressive**: It should allow all types of metadata associated with a data product to be represented in a clear and unambiguous way.

- **Human- and machine-readable**: It should be usable by platform services and external tools but at the same time be easily readable by people involved in the development or use of the product.

- **Flexible**: It should allow metadata to be defined at different levels of detail. This way, it is possible to require more or less information from the development team depending on the strategic classification of the product they are developing.

- **Extensible**: It should be easily extensible to add new types of metadata if needed in the future.

- **Composable**: It should allow referencing external standards where possible and potentially already in use. There is no point reinventing the wheel in these cases. If there are already open standards for representing a type of metadata, it is essential that the specification allows it to be referenced and does not force the use of its custom format.

- **Technology independent**: It should be as independent as possible from the underlying technology specifications, maintaining a declarative structure that limits implementation details. This way, the same specification can be used in multiple application contexts. At the same time, the more independent the specification is from the technology, the more resilient it can be to its variation over time.

When defining the specification, it is important to proceed in steps following an incremental approach. It does not make sense to require development teams to provide metadata that will not be used. It is better to start with a minimal specification that contains only strictly necessary metadata and then extend it over time as the need for it grows, depending on the level of maturity in the adoption of the data product-centric paradigm. Ideally, each metadata element requested should have a demonstrably useful purpose. It should be possible to articulate how it will be used and the value it will contribute.

Once the specification has been defined and shared, all product teams will have to take care to produce a descriptor file compliant with it during development. At the same time, the platform capability should provide the tools to store and validate the descriptor documents during the deployment of the products. It should not be possible to deploy products without descriptor files or with invalid descriptor files. The platform should also allow searching for products using the information present in the stored and indexed descriptor files. Finally, the platform should allow the distribution of the metadata present in the descriptor files to interested external applications (for example data catalog, data marketplace, and so on).

We will discuss the basic services that the platform must offer and the possible implementation methods in *Chapter 7, Automating Data Product Lifecycle Management*. In the next section, we will see an example of a specification for describing a data product using a descriptor document.

Introducing the Data Product Descriptor Specification

Multiple specifications are available for defining a data product descriptor document. Unfortunately, none of them are yet widely adopted and supported by the ecosystem of tools that provide features for building data management solutions. Therefore, there is no de facto standard to rely on, and it doesn't seem like there will be one anytime soon. This is obviously unfortunate, as the presence of a standard would reduce integration costs between different products and facilitate data product sharing and data product monetization between organizations.

It is still possible to start with one of the existing specifications and use it as is or adapt it to your needs. Some of the existing specifications focus on defining the data contract, neglecting or ignoring altogether the metadata related to the internal components of a data product. Others have a more balanced approach. Often, specifications that contain detailed metadata related to the internal components of a data product are defined to support specific governance or DataOps platforms. Sometimes, these specifications can be strongly tied to the features exposed by the platform that guided their definition.

Providing a comparative analysis of existing specifications is beyond the scope of this section. Some of the most popular specifications are listed in the *Further reading* section at the end of the chapter. When choosing whether to adopt one and, if so, which one, we suggest using the six principles listed in the previous section to make your assessment.

As an example of a specification, we will use the **Data Product Descriptor Specification** or **DPDS** for short (https://dpds.opendatamesh.org/) defined within the **Open Data Mesh Initiative** or **ODM Initiative** for short (https://initiative.opendatamesh.org/). This specification, to which I had the opportunity to personally contribute, allows you to manage both metadata related to the data product interfaces and metadata related to its internal components. It is an open specification that anyone can contribute to and respects the six principles listed in the previous section. Structurally, it is similar to the *OpenAPI specification*. Its understanding and adoption should be facilitated for software and data engineers who already use OpenAPI. DPDS is finally supported by the **Open Data Mesh Platform** or **ODM Platform** for short (`https://platform.opendatamesh.org/`), an open source platform that provides various services to automate the management of the data product life cycle.

In DPDS, a data product descriptor is a document encoded in JSON or YAML and stored for convenience on one or more files. The file containing the root of the document is the entry point that then references any other relevant files. The specification defines the structure of the document, describing its components and, for each of them, the attributes that characterize it. The following diagram shows the top-level components that make up a data product descriptor represented by means of DPDS (*Figure 5.8*):

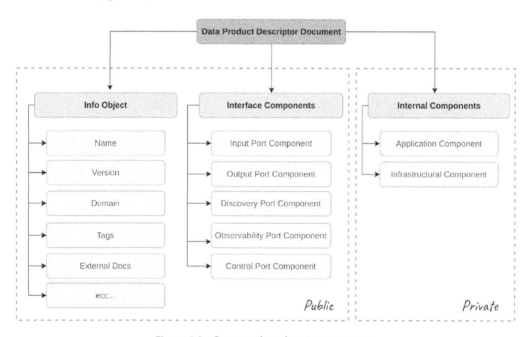

Figure 5.8 – Data product descriptor structure

Each data product descriptor in DPDS has a component called `Info` Object that contains general information about the data product, such as name, version, domain, and so on. Aside from that, there are the interface components that describe the ports. There can be multiple ports for each type of port. For example, a data product can have more than one outport port, and the same applies to any other type of port. Finally, there are the internal components that are divided into application components and infrastructure components. The former describe the applications that make up the data pipelines used by the product to transform and enrich the managed data, and the latter the infrastructure resources used by such applications. The following code block shows the basic structure of a data product descriptor:

```json
{
    "dataProductDescriptor": "1.0.0",
    "info": {
        "name": "inventory"
        "version": "1.5.0",
        "domain": "Inventory Control",
        "owner": {
            "id": "andrea.gioia@lux.com",
            "name": "Andrea Gioia"
        }
    },
    "interfaceComponents": {
        "inputPorts": [],
        "outputPorts": []
    },
    "internalComponents": {
        "applicationComponents": [],
        "infrastructuralComponents": []
    }
}
```

In the next section, we will see how to define ports within a data product descriptor using DPDS.

Describing interface components in DPDS

All ports of a data product are described within the `interfaceComponents` Object referenced by the `interfaceComponents` property of the descriptor file. This object contains the description of each port defined for the specific data product. The descriptions of the ports are divided by the type of port. Each type of port can have specific properties. However, since each port is a service that the product offers externally, the description of each port, regardless of the type, shares a basic structure and key properties. In particular, each port, in addition to having a unique name and a version, groups its properties into the following three main blocks: promises, expectations, and obligations. This way of structuring the information related to the service agreements between the data product and its consumers is influenced by the promise theory defined by Mark Burgess. The *Further reading* section at the end of this chapter provides resources for those interested in learning more about this topic.

The `Promises` Object contains all the metadata through which the data product declares the intent of the port. The term "promises" indicates that what is declared should not be taken by the consumer as a guarantee of a result but only as a commitment of the data product to behave in a way that is consistent with the promises made in order to keep them. Obviously, the more a data product keeps its promises, the more trustworthy it is. The more trustworthy a data product is, the more likely it is that a larger number of potential consumers will use it. The trustworthiness of a data product is measured based on the product's ability to keep its promises over time. This measurement should be carried out by the self serve platform and made available to potential consumers. The structure of the promises block is shown in the following table taken directly from the DPDS documentation (*Figure 5.9*):

Field Name	Type	Description
platform	string	The target technological platform in which the services associated with the given port operate. It contains usually the infrastructure provider and data center location. Optionally it can contains also the specific runtime technology used. Examples: `onprem:milan-1`, `aws:eu-south-1`, `aws:eu-south-1:redshift`.
servicesType	string	The type of services associated with the given port. Examples: `soap-services`, `rest-services`, `odata-services`, `streaming-services`, `datastore-services`.
api	Standard Definition Object	The formal description of port's services API. It is RECOMMENDED to use Open API Specification for restful services, Async API Specification for streaming services and DataStore API Specification for data store connection based services. Other specifications MAY be used as required.
depreceationPolicy	Standard Definition Object	The deprecation policy adopted by the port. It is RECOMMENDED to specify at least how long the deprecation period will be after the release of a new major version.
slo	Standard Definition Object	The service level objectives supported by the port. It is RECOMMENDED to group SLO by category (ex. operational SLO, quality SLO, etc ...) and specify them in an easy to compute way.

Figure 5.9 – Promises object structure

The `platform` and `serviceType` properties are of type `string`, while `api`, `deprecationPolicy`, and `slo` are objects of type `Standard Definition`. In DPDS, the `Standard Definition` Object is used to reference an external specification used to describe a component of the data product. For example, DPDS does not specify how to define an API, a deprecation policy, or a **service-level objective** (**SLO**), but it allows you to reference external standards or custom specifications to do so. For APIs, it suggests, but does not require, the adoption of the **OpenAPI** standard for ports that expose restful services, **AsyncAPI** for ports that expose streaming services, and **Datastore API** for ports that expose services for accessing queryable datastores. The structure of the `Standard Definition` `Object` is shown in the following table taken directly from the DPDS documentation (*Figure 5.10*):

Field Name	Type	Description
id	string:uuid	**(READONLY)** It's an UUID version 3 (see RFC-4122) generated server side during data product creation as SHA-1 hash of the port's `fullyQualifiedName`. It MAY be used when calling the API exposed by the `data product experience plane` to referentiate the component. Because the `fullyQualifiedName` is globally unique also the `id` is globally unique, anyway to referentiate the data product when calling API different from the ones exposed by the `data product experience plane` the component's `fullyQualifiedName` MUST be always used. Example: `"id": "3235744b-8d2e-57b5-afba-f66862cc6a21"`
fullyQualifiedName	string:fqn	**(READONLY)**. The unique universal idetifier of the component. It MUST be a URN of the form `urn:dpds:{mesh-namespace}:{entity-type}s:{name}:{version}`. Example: `"fullyQualifiedName": "urn:dpds:it.quantyca:apis:customApi:1"`.
entityType	string:alphanumeric	**(READONLY)** The type of the entity (ex. api, template, ecc...)
name	string:name	**(REQUIRED)** The name of the component. It MUST be unique within the component of the same type. It's RECOMMENDED to use a cammel case formatted string. Example `"name": "tmsTripCDC"`.
version	string:version	**(REQUIRED)** The semantic version number of the component.
displayName	string	The human readable name of the component. It SHOULD be used by frontend tool to visualize component's name in place of the `name` property. It's RECOMMENDED to not use the same `displayName` for different components of the same type.
description	string	The object descripion. CommonMark syntax MAY be used for rich text representation.
specification	string	**(REQUIRED)** The external specification used in the `definition`.
specificationVersion	string	The version of the external specification used in the `definition`. If not defined the version MUST be included in the definition itself.
definition	object | string | Reference Object	**(REQUIRED)** The formal definition built using the spcification declared in the `specification` field.

Figure 5.10 – Standard Definition object structure

The following code block shows the definition of the output ports of the **Inventory Data Product**:

```json
{
    "outputPorts": [{
        "name": "inventory-events",
        "version": "1.0.0",
        "description": "A streaming service that exposes delta stock
         events",
        "specification": "asyncapi",
        "specificationVersion": "3.0.0",
        "definition": {
        "mediaType": "text/yaml",
        "$href": "https://github.com/lux/inventory/blob/main/api/
          asyncapi.yaml"
    }, {
        "name": "inventory-status",
        "version": "1.0.0",
```

```
        "description": "A REST service to get the inventory status of a
         product",
        "specification": "openapi",
        "specificationVersion": "3.1.0",
        "definition": {
        "mediaType": "text/json",
        "$href": "https://api.lux.com/inventory/v1/openapi.json"
    }]
  }
```

The Expectations Object contains all the metadata through which the data product declares how the port should be used by consumers. Expectations are the opposite of promises. They are a way to explicitly declare the promises that the data product would like consumers to make about how they will use the port. The structure of the Expectations Object is shown in the following table taken directly from the DPDS documentation (*Figure 5.11*):

Field Name	Type	Description
audience	Standard Definition Object	The audience of consumers for whom the the port has been designed. It is RECOMMENDED to specify inclusion and exclusion criteria in a way that is not ambiguous.
usage	Standard Definition Object	The usage patterns for which the port has been designed.

Figure 5.11 – Expectations object structure

The Obligations Object finally contains all the metadata through which the data product declares the promises and expectations that must be respected by both it and its consumers in order to avoid penalties such as monetary sanctions or service interruption. The structure of the Obligations Object is shown in the following table taken directly from the DPDS documentation (*Figure 5.12*):

Field Name	Type	Description
termsAndConditions	Standard Definition Object	The terms and conditions defined on the port on which consumers must agree on and respect in order to use it.
billingPolicy	Standard Definition Object	The billing policy defined on the port on which consumers must agree on and respect in order to use it.
sla	Standard Definition Object	The service level agreements supported by the port. It is RECOMMENDED to group SLA by category (ex. operational SLA, quality SLA, ecc ...) and specify them in an easy to compute way.

Figure 5.12 – Obligations object structure

We will see in the next section how to describe the internal components of a data product.

Describing internal components in DPDS

The internal components of a data product can be applications or infrastructure resources.

Applications implement the logic by which data is acquired through input ports, transformed through data pipelines, and exposed to consumers through output ports. Applications can also implement functions useful for the operational management of the data product, including implementing services exposed through control, observability, and discoverability ports. The following code block shows the definition of the `Application Components Object` of the *Inventory Data Product*:

```
{
    "applicationComponents": [{
        "name": "ingestion-app",
        "version": "1.0.1",
        "description": "The sidecar app that read data through input
         ports ",
        "platform": "aws:eu-south-1:eks",
        "applicationType":"sourcing-sidecar"
    }, {
        "name": "event-processor-app",
        "version": "1.1.0",
        "description": "The app that transform input data into delta
         stock events",
        "platform": "aws:eu-south-1:confluent",
        "applicationType":"stream-transformation"
    }, {
        "name": "db-sink-connector-app",
        "version": "1.1.3",
        "description": "The app that stores the inventory status in a
         postgresdb",
        "platform": "aws:eu-south-1:confluent",
        "applicationType":"stream-transformation"
    }, {
        "name": "consumption-app",
        "version": "1.0.1",
        "description": "The sidecar app that exposes data through
         output ports,
        "platform": "aws:eu-south-1:eks",
        "applicationType":"consumption-sidecar"
    }]
}
```

Applications use various infrastructure resources to operate (compute) and store processed data (storage). The following code block shows the definition of the `Infrastructural Components Object` of the *Inventory Data Product*:

```
{
    "infrastructuralComponents": [{
        "name": "eventStore",
        "version": "1.0.1",
        "description": "The kafka topics topology required to store
          events offloaded from point of sales and the domain events
          generated by eventProcessor application",
        "platform": "aws:eu-south-1:confluent",
        "infrastructureType":"storage-resource"
    }, {
        "name": "stateStore",
        "version": "1.0.0",
        "description": "The database used by dbSinkConnector application
          to store the inventory updated state",
        "platform": "aws:eu-south-1:postgres",
        "infrastructureType":"storage-resource"
    }, {
        "name": "computeEngine",
        "version": "1.0.0",
        "description": "The cluster used by dbSinkConnector application
          to process the Inventory updated state",
        "platform": "aws:eu-south-1:eks",
        "infrastructureType":"compute-resource"
    }]
}
```

Information related to the life cycle of applications and infrastructural components can also be associated with `Internal Components Object`. This information is useful for the platform to manage internal components' provisioning or deployment. We will explore this part of the data product descriptor in more detail in *Chapter 7, Automating Data Product Lifecycle Management*.

In general, there are many more attributes that can be associated with the various components of a data product descriptor in DPDS than those shown in the code examples provided in this section. Some are included in the complete *Inventory Data Product* descriptor file, which is available in the Git repository associated with this book. However, to get a complete picture of the possible attributes that can be associated with the various components of a data descriptor in DPDS, it is advisable to refer directly to the specification documentation.

In this section, we have seen how to manage the metadata associated with a data product. In the next section, we will look at the key elements to consider when managing the data processed by a data product.

Managing data product data

Up to this point, we've explored how to identify data products to implement in order to support business cases, define the data each product manages and how it's exposed, and finally, how to formally describe them using a data product descriptor. Let's now see what are the points of attention to consider in the development of the internal components of a data product that deal with sourcing, processing, and serving data.

Sourcing data

The data product canvas allows you to determine which data each product needs, where it intends to retrieve it from, and how. All of this information can be formally described in the descriptor file in the section dedicated to input ports.

Each input port can acquire data of interest from an external system in the case of source-aligned data products or from another data product in all other cases. Regardless of the source, the data of interest can be communicated from the source to the data product (push mode) or directly acquired by the data product by reading from the source (pop mode).

In the first case, the service that the source must invoke to pass the data is described using the `api` attribute of the `Promises Object`. In the second case, the `api` attribute is not set, but within the `Promises Object`, is possible to specify which APIs of the source will be invoked (for example, which endpoint) and in what mode (for example, frequency, parameters, query, and so on) to read the data of interest.

In both cases, the `Expectation Object` can be used to indicate the constraints on the acquired data that must be verified in order for the data product to process them correctly. If the data product reads from the output port of another data product, the expectations should indicate the subset of promises made by the upstream data product that are relevant to its operation. It does not make sense to have expectations that go beyond what the source wants or is able to promise.

Once the method for acquiring data from input ports has been declared in the data product descriptor, it is necessary to implement the services for all data that will be acquired in push mode and the processes that will be executed for all data that will be acquired in pop mode. These are application components that are responsible for acquiring input data and saving it in the appropriate ingestion datastores so that it can be processed. In **hexagonal architectures**, they are called **adapters**, and they isolate the internal business logic from the way data is acquired from the outside. If the data exposure methods of the source change, there is no impact on the transformation logic, only on the adapter.

An adapter developed to acquire data through a specific API, whether exposed by the source (push mode) or by the product itself (pull mode), should take care of not only the interaction with the data source but also deciding which data to actually acquire, when to acquire it, and finally how to transform it once acquired to make it ready for processing.

The information about which data to acquire is formalized in the port description within the data product descriptor and is aligned with or even drives the behavior of the adapter. Within the Promises Object, it's possible to specify the parameters or even the query to use to filter the input data and consider only a subset of it. It is useful to filter out the data not used to reduce dependency on the source. Additionally, the input data can be filtered to read from the source only the data that has actually changed since the last read (delta load).

Within the Promises Object, it is also possible to specify when to trigger the data acquisition. Even when data is pushed by the source, it may make sense to define triggering policies. The data product may decide not to always acquire the data that the source pushes, but only at certain frequencies or when certain events occur.

In general, there are two types of triggering policies: time-based triggers and event-based triggers. Time-based triggers read the data at a specified frequency. Event-based triggers read the data only when specific conditions occur. In this case, the adapter must be able to receive event notifications from the outside (for example, from the self serve platform) in order to decide if the desired conditions have been met to proceed with new data acquisition from the source.

The information on how to transform the acquired data to make it available to the transformation pipelines is formalized in the description of the application component associated with the adapter within the descriptor. The adapter can use it as configurations to decide which transformations to apply once the data is acquired before saving it in an ingestion datastore.

The transformations performed by adapters do not contain business logic. In other words, they do not enrich the input data. This is the task of the data pipelines. Adapters perform just technical transformations such as data decryption or transformation of the structure or model used to represent the data in a way that conforms to the expectations of the data pipelines that will then enrich it (ACL).

In general, it is recommended to adopt an event-triggered pop acquisition mode when implementing an input port. This reduces the dependency between data products since the source does not need to know its consumers and is not responsible for passing data directly to them. The source only needs to communicate its state changes to the system through its observability ports (for example, when exposed data is updated).

The data product consumer can directly read the changes from the observability port of the product it wants to consume from to evaluate whether to acquire the data. Alternatively, this activity can be done by a notification system offered by the self serve platform that allows for even greater decoupling of the producer from the consumer. In any case, regardless of how the interaction between producer and consumer takes place, it should never be orchestrated externally. It is the responsibility of the consumer to decide when and what to consume, not an external orchestrator, as is usually the case in monolithic data management solutions.

When managing data through data products, the goal is for each product to be self-sufficient and independent. To achieve this, a distributed coordination pattern is preferred, similar to microservices architectures. This pattern, called choreography, relies on individual components understanding their roles and interacting with each other directly rather than relying on a central orchestrator.

To further streamline development, it is suggested to identify common functionalities across adapters and build reusable components at the platform level. These pre-built building blocks can then be incorporated into various data products, saving development time. We will explore the topic of building blocks in *Chapter 7, Automating Data Product Lifecycle Management*. Here, we will simply observe that when the development of the adapters used by product teams to acquire data through input ports is centralized at the platform level, it also becomes possible to instrument the adapters with different traffic monitoring, routing, security, and cost allocation capabilities.

Once the data of interest has been acquired, it needs to be processed. Let's see how in the next section.

Processing data

The data product canvas allows you to determine which business logic the data product needs to apply to the input data to enrich it and make it ready for consumption. To transform the input data in a way that is consistent with the purpose for which the product has been defined, it is necessary to implement one or more applications organized into one or more data pipelines.

The applications that make up the data pipelines use various infrastructure resources, both in terms of compute and storage. The way in which data pipelines are implemented, and consequently the infrastructure resources used, depends heavily on the integration framework chosen by the development team.

Since the implementation of business logic is not visible to consumers, each development team is theoretically free to choose the integration framework they prefer independently. In practice, however, it is advisable to identify some reference frameworks for managing the main integration methods in the various supported environments. Platform teams focus on developing building blocks, blueprints, and services to support these frameworks, making them easier for product teams to use.

While platform teams provide reference frameworks and components to streamline development, product teams should retain the flexibility to choose alternative integration tools and infrastructure resources. This autonomy is crucial when specific needs arise, but it requires the product team to have the necessary resources for ongoing management of the tools and resources chosen without relying on the platform team.

At the infrastructure level, reference components can be created within a dedicated platform instance for each new product. For example, a data product might have its own dedicated instance of an analytical database. Alternatively, there can be a single central instance of the platform, managed by the platform team at an operational level. Within this central instance, each data product has some resources created within a dedicated tenant. For example, a data product might have dedicated schemas within a shared instance of an analytical database. The two options are complementary, and the choice of which to adopt generally depends on the product being developed, its workload segregation requirements, and its security requirements.

The same considerations made for infrastructure components apply to reference integration frameworks. Some products might have their own instance of the integration framework, while others might only have a dedicated tenant within a central and shared instance of the framework.

The way in which data pipelines are implemented depends heavily on the integration frameworks and infrastructure resources that the development team decides to use. In addition to developing the pipelines, the team is also responsible for testing, releasing, and managing them in production, ensuring the reproducibility and predictability of the results.

Reproducibility of results aims to guarantee consumers that readings of the same data at two different times will not produce different results. Reproducibility is achieved through the immutability of managed data. Once acquired and exposed to consumers, immutable data cannot be modified:

- If the data is an event (for example, stock variation), it is immutable by nature. New events are acquired and added to the previous ones. If an event is incomplete or incorrect, it is no longer modified. However, the modification of the state of the entity to which the event refers (for example, inventory) can be rectified by subsequent events.

- If the data is a state (for example, the state of the stock of product X at 2 P.M.), guaranteeing immutability requires a more articulated management of the time dimension. It is necessary to keep track of both the validity time of the data and the transaction time (**bi-temporality**).

Validity time refers to the time at which the data can be considered correct in the real world, while **transaction time** refers to the time at which the data was recorded in the system. With this dual time management, it is possible to guarantee the immutability of the data even when it is a state and not an event.

The state of the stock at 2 P.M. (validity time) recorded in the system at 3 P.M. (transaction time) can be different from the state of the stock recorded in the system at 4 P.M. (transaction time). New data read between 3 P.M. and 4 P.M. could have led to a correction of the correct state of the stock at 2 P.M. However, the state at 2 P.M. known to the system at 3 P.M. does not change.

A consumer interested only in having the most updated version of the inventory state at 2 P.M. can read inventory data with the most recent transaction time related to the validity time, 2 P.M. Repeating this reading at two different times, the values obtained can obviously change. However, a consumer can always specify a specific transaction time in addition to the validity time. In this case, the outcome of the readings does not change; the data will always be the same.

There are two main approaches to managing bitemporality when dealing with data related to the state of an entity: materialization and snapshotting. Materialization reconstructs the state of the data at a specific point in time (transaction time) by processing historical events. This is ideal when historical events are readily available. Snapshotting creates periodic "snapshots" of the data state, each associated with a specific transaction time. This is a good alternative when event history is unavailable or materialization is too slow for performance needs.

Most modern storage systems available today mask the complexity of managing data bitemporality through time travel functionality. Time travel functionality allows delegating to the system the management of data along the transaction time dimension. When a new version of old data arrives, an update can be safely made to update the value related to the specific validity time. The system creates a new version of the data behind the scenes. If then it is needed to trace the value of the data at a specific transaction time, it's possible to ask the system to recall that version of the data without worrying about how it implements the data versioning functionality internally.

Data retention policies indicate how far back in time you can go to retrieve the value of data at a specific transaction time and the granularity of these steps. For example, the stock could have a retention policy that allows consumers to retrieve the inventory state at any transaction time within the last 2 days. Instead, data retention for transaction times exceeding 2 days is limited to snapshots captured at varying frequencies. Daily snapshots are available for the preceding year. Monthly snapshots are provided for data between 1 and 2 years prior. Beyond 2 years, only annual snapshots are retained.

Retention policies must be specified inside the descriptor in the `Promises Object` of the output port that exposes the data.

Predictable results are essential for consumers to trust the data they read. Reproducibility is achieved through idempotence, versioning, and isolation of developed pipelines.

Idempotence ensures that a pipeline always produces the same output data when reading the same input data. It also guarantees that no duplicate results are generated when the same input data is read multiple times.

Versioning of pipelines ensures that even when the business logic changes, it is always possible to trace back to the version of the pipeline that generated specific data. If necessary, a specific version of the business logic can be reproduced with new input data.

Isolation aims to reduce dependencies on the execution context, which could generate undesired side effects. An isolated data pipeline should write its data to dedicated datastore that are not shared with other data pipelines. It should also be able to intercept errors in the input data in order to correct them or, at least, not write incorrect results to the output data stores.

In the next section, we'll explore how to expose processed data for consumption.

Serving data

The data product canvas helps determine how a data product should expose enriched data. Each method supports a specific type of consumption and is associated with an output port. The more consumption methods offered, the more versatile and reusable the product is for supporting multiple business cases.

A data product can be consumed in various ways to suit different user needs. Users can perform analytical queries on the data through online querying methods. For specific data point retrieval, punctual access methods are used. Additionally, data can be consumed as a continuous stream of events using streaming access methods.

DPDS recommends specific APIs for each consumption method:

- Datastore API for online querying
- OpenAPI for punctual access
- AsyncAPI for streaming access

Similar to input ports, output ports should be implemented using dedicated adapters. These adapters transform the output data according to the needs of the consumption method, implement the consumption service, and control security and access policies.

When transferring data between products, the adapters used for output and input ports can communicate directly. Alternatively, a platform-level transport layer can mediate the communication. If a transport layer is present, it defines data broker tools and policies for routing and securing data during transport. For example, the transport layer might rely on a data virtualization or data sharing platform for all at-rest data, and a streaming platform for all data exchanged through real-time events. Routing and security logic can then be developed on top of these data broker tools.

As with input port adapters, output port adapters can be progressively standardized and managed by a central platform team.

Summary

In this chapter, we explored the design and development of a data product.

We began by examining how to define the interactions between a data product and its external environment, both locally within the bounded context and globally within the entire organization ecosystem. Correctly designing these interactions is crucial because it ensures interoperability between products and effective coordination between the teams that develop them.

Next, we examined the main internal components of a data product and how to describe them using a descriptor file. Defining a descriptor file is essential for uniformly managing the metadata related to data products, thus facilitating their discovery and understanding by consumers and their life cycle automation by the self serve platform. The main characteristics of a good specification for defining a descriptor file are expressiveness, computability, flexibility, extensibility, composability, and independence from specific technologies. We used DPDS to show an example of specification.

Finally, we explored how to develop the internal components of a data product that acquire, transform, and expose data. Since these are internal components, we did not define prescriptive implementation methods. Teams are free to choose the methodologies and tools they prefer. However, we outlined some key principles to follow in their implementation in order to make the data assets managed by data products easier for development teams to manage and easier for potential consumers to use.

In the next chapter, we will see how to manage the operation of a data product in a production environment.

Further reading

For more information on the topics covered in this chapter, please see the following resources:

- *Context Mapping* – DDD-Crew `https://github.com/ddd-crew/context-mapping`

- *Building an Event-Driven Data Mesh* – Adam Bellemare (2023) `https://www.amazon.com/Building-Event-Driven-Data-Mesh-Architectures/dp/1098127609`

- *What is a Composable Business Application?* – Massimo Pezzini (2023) `https://www.youtube.com/watch?v=_oHqiDWTrAQ&t=1333s`

- *Data Product Descriptor Specification* – Open Data Mesh Initiative `https://dpds.opendatamesh.org/`

- *Data Product Specification* – AgileLab `https://github.com/agile-lab-dev/Data-Product-Specification`

- *Open Data Contract Standard* – BITOL `https://github.com/bitol-io/open-data-contract-standard`

- *Functional Data Engineering — a modern paradigm for batch data processing* – Maxime Beauchemin (2018) `https://maximebeauchemin.medium.com/functional-data-engineering-a-modern-paradigm-for-batch-data-processing-2327ec32c42a`

- *Thinking in Promises: Designing Systems for Cooperation* – M. Burgess (2015) `https://www.amazon.com/Thinking-Promises-Designing-Systems-Cooperation/dp/1491917873`

6

Operating Data Products in Production

In the previous chapter, we explored the design and implementation of data products. This chapter delves into their release and governance in production environments.

We will delve into release methodologies by introducing the principles of **Continuous Integration and Continuous Delivery (CI/CD)**. We will explore the benefits of these approaches and look at how to build an automated release pipeline to support them.

Next, we will explore the governance of data product instances in production environments. This encompasses utilizing observability and control ports, along with computational policies, to ensure data product integrity and adherence to organizational standards.

Finally, we will see how to manage the consumption of data exposed by data products. We will see how to make the data products easily searchable and accessible. We will then show how to compose them together to support multiple business cases and how to manage their evolution over time. By mastering these techniques, you'll ensure your data products are released smoothly, operate reliably, and deliver ongoing value to your organization.

This chapter will cover the following main topics:

- Deploying data products
- Governing data products
- Consuming data products
- Evolving data products

Deploying data products

A data product is the unit of modularization of the data architecture. As such, it self-contains all the components necessary for its operation and must always be deployable in a runtime environment in an atomic and autonomous way. In this section, we will see how to define the deployment pipeline of a data product by adapting the practices of CI/CD used widely and successfully in the world of software development to the data world.

Understanding continuous integration

As seen in previous chapters, a data product consists of the data it exposes; the applications that acquire, transform, and share it; the infrastructural components that serve the applications; and finally, all the metadata that describes the previous components, usually collected in a descriptor file.

During the development phase, the applications are implemented, the infrastructural components they will use are defined, and the descriptor file is populated according to the specifications defined by the governance capability. The data itself is not an object directly managed during the development of the data product. It is the applications that manage the data once the data product is deployed in a runtime environment.

The development team should be actively involved in, if not directly responsible for, deploying the data product to the runtime environment. Their involvement in the release process fosters greater ownership and faster issue resolution, resulting in more reliable, consistent, and agile deployments.

Once all the product's components are developed, it's essential to first ensure they integrate correctly and produce the expected results as specified in the descriptor file before releasing the data product into a production environment.

However, performing the integration process only once, when the development of all the components is finished (**pre-release integration**), is risky. When the product components are developed in parallel by different team members for long periods, it is likely to have multiple integration problems at the end. Each problem requires rework, which can potentially be time-consuming. Since the number and complexity of integration problems present is not known until the end of development, it is difficult to predict how long the integration phase will last. Consequently, this way of proceeding makes the release times of new data product versions unpredictable.

For this reason, modern approaches to software development prefer to increase the frequency with which the integration between the components is verified. The more frequent the checks are, the more readily integration issues can be detected and addressed before the discrepancies between interconnected components grow too significant, requiring extensive time and effort to rectify.

In particular, the practice of **CI** requires that the integration between the components is verified at least once a day, and if possible, even multiple times a day based on the actual changes made to the components. Let's see how in the next section.

Automating the build and release processes

A data product can be viewed as an application that processes and exposes data. Therefore, the CI practices that streamline transactional application development can be equally beneficial for data products. The following diagram (*Figure 6.1*) outlines the various stages involved in validating the integrability of a data product's components.

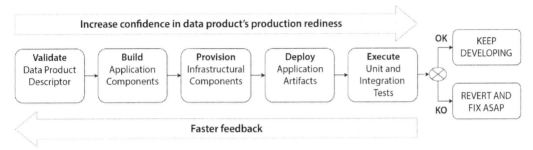

Figure 6.1 – CI

To ensure continuous flow, as requested by CI practice, development team members must share their changes at least once a day rather than waiting for the completion of their assigned development tasks.

A **Version Control System** (**VCS**) such as Git is crucial for sharing and tracking changes effectively. The descriptor file and all products' components, both application and infrastructural, must be placed under version control.

It is possible to manage versioning for all components of a data product through a single repository, or to have repositories that group components according to different criteria (such as type, module, technology used, etc.). However, every data product must have a main repository that contains the descriptor file. The descriptor file then references the repository where each component is versioned, if it is different from the main one.

For application components, pipelines, and adapters, the source code is typically versioned. When using external frameworks (especially integration tools) for development, the versioning methods for the developed objects may vary. Not all frameworks integrate seamlessly with traditional code versioning systems. Some have their own versioning system, while others do not manage versioning at all.

Regardless of the framework used, it is essential to define an operational mode that allows us to isolate the development of new features (**feature branch**) from the main version (**main branch**). The main branch is used during the integration process. It is critical that intermediate developments are not made directly on the main branch to avoid compromising the integration process.

Every time a significant modification is completed, it must be merged into the main branch, triggering the integration process. To maintain a high frequency of integration testing, it's also crucial to break down work into small, manageable units, avoiding long-lived feature branches. This approach prevents large, infrequent merges and ensures that integration, testing, and releasing occur continuously and efficiently.

The operational framework model must define how to merge development and main versions, as well as how to initiate the integration process whenever the main version is modified. These operational tasks can more or less be automated depending on the level of versioning support provided by the framework used.

Infrastructure as code

Infrastructure as Code (IaC) replaces manual infrastructure configuration with automated scripting. Instead of clicking through complex interfaces, IaC lets you define your infrastructure's desired state using code. This code specifies details such as the number of servers, storage capacity, and network configurations. IaC tools then translate this code into instructions for creating and configuring those resources automatically. This approach offers several benefits; **Automation** in which IaC automates infrastructure provisioning, saving time and reducing errors, **Repeatability** in which IaC scripts allow for consistent deployments across different environments and easy rebuilds if needed, and the final benefit is **Version control** where IaC code can be version controlled, enabling change tracking, rollbacks, and collaboration.

By offering a more efficient and reliable approach to infrastructure management, IaC is becoming a valuable tool for many organizations.

Infrastructure components should be defined using an IaC tool, such as **Terraform**, to enable automatic, secure, and reproducible provisioning when needed. The declarative definition of an infrastructure component as code can be versioned using the same version control system used for the application code.

If an infrastructure resource cannot be declaratively defined as code, it is recommended to write an ad-hoc application to manage its provisioning. In this case, the application code is versioned.

Once triggered, the first step of the integration process is validating the descriptor file (*validation phase*). Since deploying a data product without a descriptor file is not possible, it's important that it be properly updated with each modification of the data product. To prevent the descriptor file from being compiled only at the end of developments, it is therefore useful to enforce the validation of this file at the beginning of each integration process. We will discuss how to manage the validation of the descriptor file in the *Implementing computational policies* section.

The validation phase is followed by the **build phase**. The goal of the build phase is to generate artifacts for each component that can then be released to a runtime environment. Not all components require a build phase. This phase is generally necessary for code-based application components that need to be compiled and packaged before they can be released and executed.

The build phase is followed by the **provisioning phase**. The goal of the provisioning phase is to create a runtime environment where the data product components can be deployed and integration tests can be executed. To provision the runtime environment, all infrastructure components that are part of the data product must be provisioned. These can be a dedicated infrastructure instance (e.g., a dedicated analytical database instance) or dedicated resources within a shared infrastructure instance (e.g., a dedicated schema within a shared analytical database instance).

The provisioning phase is followed by the **deployment phase**. The goal of the deployment phase is to ensure that all application components are properly operational in the runtime environment provisioned for testing. This requires the deployment of the components or their associated artifacts, generated during the build phase, and their correct configuration.

The deployment phase is followed by the **test execution phase**. The goal of this phase is to ensure that the application components function correctly both as individual units and as an integrated whole. To this end, specific tests for the individual application components (**unit tests**) are first executed. These tests should already have been executed by the developers of each component within their respective local development environments before starting the integration process and then resecured again during the build phase. However, it is also useful to run unit tests in the test environment, which is usually more aligned with the production environment. Let's see how in the next section.

Automating integration tests

Once the unit tests are completed, **integration tests** can be performed to verify that the various components, in addition to functioning correctly independently, also correctly combine to produce the desired final behavior. The implementation of unit tests depends strictly on the technology stack used for the implementation of the component (**white box testing**). Integration tests, on the other hand, focus on the end-to-end behavior of the data product and are therefore independent of the implementation details of the individual components (**black box testing**).

Generally, integration tests can be performed in each data product by loading test data into the landing data stores and then verifying the data generated by the product pipelines within the consumption data stores. An integration test, as shown in the following figure (*Figure 6.2*), consists therefore of a set of test data (**fixture data**) to be provided as input to the data product and a set of conditions to be verified on the data generated as output by the product (**expectations**).

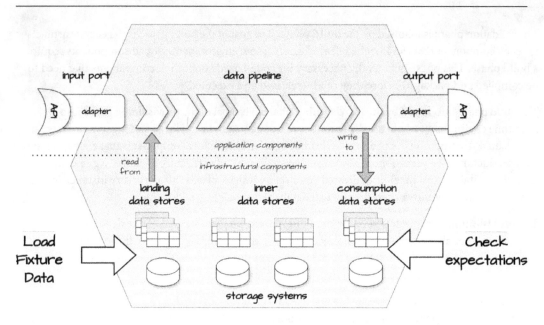

Figure 6.2 – Integration test

Integration tests can verify the end-to-end behavior of the data enrichment process or specific portions of it. The procedure remains the same: test data is loaded into the data stores from which the portion of the process being tested reads the data, and then the assumptions on the generated data are verified in the data stores where it writes its results.

A dedicated application can be developed to execute integration tests. This application can manage the following:

- Loading of test data (**test warmup**)
- Verification of assumptions on output data (**test execution**)
- Collection of results (**test result aggregation**)
- Cleaning of the environment to proceed with subsequent tests (**test cleanup**)

Since this application is generally common to each data product, its implementation can be delegated to the platform capability.

Integration tests (input fixture data and output assumptions) are an integral part of the product and, as such, should be versioned along with the other components. While unit tests are versioned in the repository that contains the component they are associated with, integration tests are generally versioned in the main repository or in a dedicated repository.

If the integration process terminates with errors, it is necessary to prioritize their resolution. When practicing CI, it is fundamental to keep the main branch of the product clean. If this does not happen, it is not possible to test new changes because the previous unfixed errors would affect the result. Additionally, developers, knowing the poor quality of the main branch, would tend not to align their local environment with it very often, increasing the probability of introducing integration problems.

Since solving an integration problem may take time, it is necessary to have a mechanism to revert the changes made to the main version to bring it back to a consistent state as soon as possible. Common version control systems allow for easy management of this **revert operation** by restoring a previous working version. For any frameworks used that do not support or only partially support version control, revert management must be handled manually or with custom solutions integrated into the development process.

If the integration process terminates successfully, it means that at the current development state, the infrastructure can be provisioned, the components can be built, they work correctly, and they integrate as expected. However, it does not mean that the currently developed version of the data product itself is ready to be deployed in production. In fact, the development may not be finished yet. Therefore, the current version of the data product present on the main branch may not yet be in its desired target state.

Only when the development is completed and the version of the data product present on the main branch has successfully passed the integration tests is it ready to be deployed in the production environment. We will see how in the next section.

Understanding CD

At the infrastructure level, the environment in which integration tests are performed should be identical to the production environment. This way, you can be reasonably sure that a data product version that has successfully passed integration tests will not have unexpected integration problems once deployed in production.

The infrastructure equality between the test and production environments also allows us to manage deployment in the two environments in a substantially identical way, except for the configurations applied in resource provisioning and application setup. Therefore, to release a new version of a data product in the production environment, it should be sufficient to execute the same tasks necessary for deployment in the test environment, passing the appropriate configurations for the production environment. No additional logic or specific operation should be necessary. These are the main steps to release a data product:

1. Before deploying a data product version in the production environment, it is necessary to validate its descriptor file first. The validation process is the same as the one applied for deployment in the test environment, but the checks performed could be more extensive for the production environment. Additionally, depending on the release policies defined by the organization, before proceeding with a production deployment, it might be necessary to execute a specific approval workflow that involves additional figures outside the development team, in addition to the automatic validation of the descriptor file.

2. Once a new version of a data product has successfully been deployed in production, it is necessary to perform some housekeeping activities on the data managed by the specific data product instance. These activities are generally not necessary in the test environment. First, it is necessary to retrieve the historical data processed by the old data product instance. If the structures of the data stores used internally to store this data have changed, it is necessary to transform the historical data to store it correctly in the new structures. These transformations, called **data migrations**, do not modify the content of the historical data. The data managed by a data product must be immutable, both within the single instance and between instances associated with successive versions of the data product. Data migrations only modify the form.

3. Once the historical data has been loaded and (if necessary) migrated to the new data store structure, it may be necessary to recompute it in order to have a version updated to the transformation logic implemented by the new version of the data product (**data backfill**). Depending on the type of change in the logic of the data product's data pipelines, to perform the backfill, it may be necessary to re-read the data directly from the sources and execute all the new pipelines on it, or to start from the data saved in internal data stores and execute only the modified portion of each pipeline. The processes that follow the data migration and data backfill activities are themselves data pipelines. However, unlike those seen so far, they are only executed once post-deployment. Nevertheless, it is essential to consider them as an integral part of the data product, versioning them and managing them like the other pipelines.

4. Once the deployment and housekeeping activities are completed, before making the data product instance available for consumption, it is usually necessary to perform all those tests that cannot be done directly in the test environment due to the lack of real data or other data products to consume from or expose data to. These types of tests performed directly in production are common because, even though the test and production environments are aligned in terms of infrastructure, they are often misaligned in terms of data and the versions of the deployed product instances.

Among the types of tests to be performed in this post-deployment phase are **non-functional tests** (e.g., performance tests, security tests, etc.) and **acceptance tests**. These tests can be performed once the data product instance is operational and reads real data from its sources. These are generally not one-shot tests, but rather tests that monitor the compliance of the data product instance's behavior for a predetermined period of time. If no errors or non-conformities are found at the end of the period, the tests can be considered passed.

Depending on the testing policies, in addition to the acceptance tests defined by the development team during the requirements formalization phase (**automated acceptance tests**), it may also be necessary to have acceptance tests carried out directly by the data product consumers (**user acceptance tests**). In this case, the timing and methods for conducting these tests before the data product instance is finally made available for consumption must be clearly defined.

As with integration tests, it is also advisable not to wait until the end of development to perform non-functional and acceptance tests. To make the development progress more transparent and the completion times predictable, it is useful to plan frequent intermediate releases in order to intercept any problems early and correct them.

In particular, the practice of CD ensures that the development version of a data product is always potentially releasable. This way, it is always possible to deploy the version of a data product in the main branch in the production environment to perform the necessary tests. This does not mean that for each release, the data product instance is then actually made accessible for consumption. This generally only happens at the end of development. The practice of CD is based on and extends that of CI. Both practices require strong collaboration between the development team and all the actors involved in the delivery process (**DevOps culture**), as well as extensive automation of all the activities involved in this process (**automated deployment pipeline**). In the next section, we will see how to automate this process.

Defining the deployment pipeline

To be able to perform releases to the test environment for integration testing and then releases to the production environment for non-functional and acceptance testing with high frequency, the release process needs to be as fast as possible.

To this end, it is necessary to divide the process into *stages*, identify the key *tasks* for each stage, and evaluate how and to what extent these can be automated. Ideally, manual tasks should be kept to a minimum, relegated mostly to parts of the process that require an explicit approval workflow.

The automated process that contains the tasks necessary to take a data product from the development environment to release in production through all intermediate stages is called a **deployment pipeline**. The deployment pipeline has a linear structure in which each stage follows the previous one, bringing the product closer and closer to the state in which it can be released for consumption. It should always be possible to execute the pipeline incrementally, one stage at a time. To reach a specific stage, all previous stages must have been successfully completed in the prescribed order.

The tasks associated with the transition between one stage and the next do not necessarily have to be sequential. They can generally be organized and orchestrated by means of a **Directed Acyclic Graph (DAG)**, thus enabling the possibility of executing multiple tasks in parallel where it makes sense. The tasks associated with a transition between two consecutive stages should use the artifacts produced by the tasks executed for the previous stage transitions and not redo work that has already been done previously. The artifacts produced at each stage transition must therefore be stored in an **artifact repository** so that they can be reused later by the tasks associated with subsequent stage transitions. The deployment pipeline can then be partially executed and then restarted without having to redo the work already done to reach the current stage.

It is extremely important, in automating the deployment pipeline, to ensure that the process executed is *idempotent* and *reproducible*. The process is idempotent when rerunning the pipeline from the beginning does not create problems due to conflicts generated as a result of previous executions. Therefore, it is important that the pipeline is capable of cleaning the environment from artifacts generated by previous executions or alternatively creating a new, completely isolated environment with each execution. The process is reproducible when rerunning it on the same version of the data product yields the same results. Therefore, it is important that the pipeline isolates the product as much as possible from its external dependencies, which may, over time, influence the predictability of the results produced by the pipeline itself. In this sense, the most important dependencies to be properly managed to isolate a data product from the external environment and make its deployments reproducible are those with pre-existing infrastructure, external libraries used by application components, and the data used in tests.

Regarding the pre-existing infrastructure, it is important that the product declares what type of infrastructure it expects to find during provisioning. In particular, the type and version of the services and tools it will use should be declared. If these change, a provisioning error will be promptly identified with the exact details of the unsatisfied dependencies. This makes it easy to isolate the problem and fix it faster. If there is no explicit declaration of the versions of the services and tools that the product uses, the pipeline could pass the provisioning phase without generating errors, only to generate unexpected behavior later. At this point, it may take longer to trace the cause of the problem.

Similarly, each application component should declare the exact version of each external library it uses so that the exact version is imported during the build or construction of the environment, rather than the latest stable version that can change over time, creating potential problems during the building phase or subsequent phases.

Finally, the data used for integration tests should be defined during the development phase and versioned with the product (fixture data) so that it does not change over time. This is obviously not possible for non-functional and acceptance tests, which use real data instead. These tests, which constitute the last stages of the deployment pipeline, are the only ones for which deviating from the reproducibility requirement makes sense.

Theoretically, each stage of the deployment pipeline can be executed in a dedicated, isolated environment built as a copy of the production environment. While this simplifies pipeline development, requiring less housekeeping, it is not particularly performant and could also be expensive.

To make the pipeline faster and consequently increase its frequency of execution, it is generally necessary to reuse the environment as well as the artifacts produced in the previous stages. In other words, the best way to make a pipeline fast is to avoid performing any work on components, both application and infrastructure, that have not been modified since the last execution. Therefore, it is generally common for multiple stages of the pipeline to be executed within the same runtime environment. It is even possible to reuse the application and infrastructure components of a product between pipeline executions if they have not been modified.

The number of environments and how they map to the activities of the different stages is an organizational choice. In practice, an integration testing environment and a production environment are more than sufficient in many cases. In some cases, it is even possible to execute integration tests in the production environment by appropriately isolating the data products under test and then making them visible and usable only after the actual release (**shadow deployment**).

Write-audit-publish pattern

The **Write-Audit-Publish** (**WAP**) pattern for data management was first introduced in 2017 by Michelle Winters at Netflix. This pattern is designed to test data integration pipelines directly within a production environment.

As the name suggests, WAP consists of three main phases:

Write: In this phase, the data to be integrated is copied and transformed into a part of the production environment that is not accessible to downstream consumers (shadow deployment).

Audit: During this phase, necessary tests are conducted to validate the accuracy and quality of the processed data.

Publish: If the data passes all tests successfully, it is then made available to potential consumers. If not, the data is discarded until the pipelines are fixed.

Advancements in technologies such as zero-copy cloning and time travel, which are increasingly supported by major storage systems, have made shadow deployments easier, faster, and more cost-effective. As a result, the WAP pattern has become a popular choice for organizations seeking a reliable way to test data pipelines.

Each component of a data product should have its own deployment pipeline. If possible, this pipeline should be defined declaratively as a configuration to be passed to a dedicated CI/CD solution (**pipeline as code**). This pipeline should be versioned in the same repository used for the component. If it is not possible to define the pipeline declaratively, then the component must implement a dedicated deployment service/script and version it. In both cases, the reference to the pipeline definition must be referenced within the descriptor file.

There must then also be a deployment pipeline associated with the entire data product to orchestrate the pipelines of the individual components and to perform cross-cutting tasks. This pipeline, whether defined as code or as a script/application, must be versioned in the main product repository that also contains the descriptor file. If the product deployment pipeline simply executes the pipelines of each component in sequence, then it is possible to not define it explicitly and infer it at deployment time from the information contained in the descriptor file.

The deployment pipelines of all components and the main product pipeline must be composed of exactly the same sequence of stages. When a stage is not significant for a specific component, then there will be no tasks associated with that stage for that component.

There are many solutions on the market for managing CI/CD pipelines. In addition to the native solutions offered by the major cloud providers (e.g., **Azure DevOps**, **AWS CodeStar**, **Google Cloud Build**, etc.), there are various commercial (e.g., **GitLab**) and open source (e.g., **Jenkins**) alternatives. These solutions, in addition to having the ability to orchestrate complex pipelines defined declaratively, offer services for managing infrastructure provisioning, configuring environments, and saving intermediate artifacts produced, as well as result notification.

Once the data product instance is deployed and officially released into the production environment, it enters its operational phase. In the next section, we will see how to manage a data product instance in production.

Governing data products

After deploying a data product instance, it requires governance throughout its operational lifespan within the production environment. In other words, it needs constant monitoring and control. This section will explore how to govern data products from their release to decommissioning.

Collecting and sharing metadata

Data product versions come with a descriptor file that stores all their relevant metadata. The governance capability defines a shared standard for the content and format of this file (example, DPDS). Data product development teams are responsible for gathering the necessary metadata and including it in the descriptor file. To ensure successful deployment in any runtime environment, the descriptor file in the product's main repository must be complete and accurate. Once a product version is instantiated, the instance needs to share its descriptor file with potential users. To facilitate this, each product instance should be easily addressable and always expose a discoverability port for sharing its descriptor file (*Figure 6.3*).

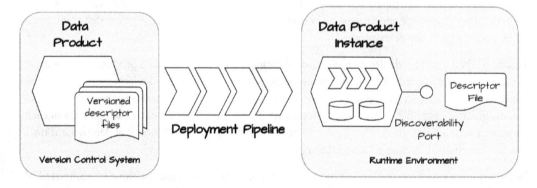

Figure 6.3 – Exposing the descriptor file through a discoverability port

Once the release of the data product is successfully completed, the deployment pipeline can notify the systems that need to receive the information contained in the descriptor, for example, through a Git Hook. These systems can then retrieve the descriptor for the specific instance via the services exposed by its discoverability port. Generally, it is sufficient to notify the self serve platform, which reads the descriptor from the observability port and then takes care of notifying all other downstream applications that require specific information about the data product contained in the descriptor.

Addressing a data product instance requires a unique URI, known as the **instance URI**. Within a specific runtime environment, potential consumers should be able to determine the instance URI based on the product's name, version, and (optionally) domain. A shared standard can define the format of the instance URI, allowing potential consumers to reconstruct it independently. For example, the instance URI could be built by concatenating the version, product name, and environment as a subdomain of the organization's main domain (e.g., `http://v1_5_0.inventory.prod. lux.com`). Alternatively, a **data product registry** managed by the platform capability can be used. During deployment, the registry records the correlation between the product's name, version, and other identifying information with its instance URI. Users can then perform lookups in the registry to find the URI of the desired product instance.

The instance URI serves as the entry point for interacting with a data product instance. It acts as the base URI for the discoverability port's service endpoints, which exposes the descriptor file (e.g., `http:// v1_5_0.inventory.prod.lux.com/descriptors/current`). Once the descriptor file is obtained, all the metadata associated with the data product instance is accessible, including the list and description of other exposed ports. The governance capability defines a standardized API for the discoverability port that exposes the descriptor. In the simplest case, this can be a REST API with a single endpoint associated with the `descriptors` collection, callable via the GET method to obtain the current instance's descriptor file. More complex versions might use additional query parameters to filter the descriptor. For example, `descriptors/current?entityType=inputport` would only return details about the input ports. Alternatively, additional endpoints could be defined to access specific components, such as `descriptors/current/outputports/inventory-events` for details on the inventory-events output port.

Some metadata within the descriptor file might depend on the environment where the product runs and only be available after deployment is complete (e.g., the hostname of the machine running a service exposed by a port). To handle this, the discoverability port service that returns the descriptor file must be able to resolve this metadata at runtime and integrate it into the descriptor exposed to users. The DPDS allows for populating the various fields of a descriptor file using variables. In the following example, the `externalDocs` property uses the `docServerUrl` variable:

```
"externalDocs": {
  "$href": https://{docServerHost}/products/inventory.html
}
```

Once the deployment process determines the variable's value, it can be passed to the product instance via a control port. The discoverability port can then substitute the variable with its actual value before returning the descriptor.

The discoverability service returns the descriptor structured according to the standard specification chosen by the governance capability (e.g., DPDS). Internal consumers within the organization and the self serve platform must be able to understand this specification. However, to facilitate access to the information describing the product instance by external applications or consumers, the discoverability port API can be extended to allow the export of the information contained in the descriptor in other formats as well. For example, it might be useful to allow reading the descriptor in **DPROD format** (`descriptors/current?format=dprod`). DPROD is an extension of the **DCAT standard** that defines an **ontology** for describing a data product. Exporting in this format can be used to expose basic data product information in public catalogs or to add product information to a **knowledge graph**. We will explore the concept of ontology and knowledge graph in more detail in *Chapter 10, Distributed Data Modeling*.

Since the descriptor contains both information of interest to consumers (e.g., port descriptions) and information of exclusive interest to the development team and the platform (e.g., application and infrastructure resource descriptions), it might be wise to filter the descriptor based on the user requesting it. Granting consumers access to sensitive details about the data product's internal components could be unnecessary and risky. This access control could also extend to port descriptions. For instance, access to some port descriptions might be restricted only to the development team or users associated with the data product's domain.

The discoverability port leverages a shared API, making its implementation consistent across all data products. Consequently, developing it as a reusable building block managed by the platform capability is a logical step. Each product can then instantiate and reuse this building block.

We've explored how to collect and share metadata describing a product instance through a dedicated discoverability port. Now, let's delve into how this metadata can be used to verify the product's compliance with policies defined by the governance capability or by the development team itself.

Implementing computational policy

In *Chapter 3, Data Product-Centered Architecture*, we explored how the governance capability defines policies that data products must follow to ensure interoperability. In simpler terms, these policies guarantee that data products can be combined in various configurations to support different business cases. Enforcing these policies is the responsibility of the control function (System 3). The platform capability should automate policy enforcement to streamline the process, leaving only the task of performing a second-level check to the control function. Policies that can be automatically verified are called **computational policies**.

A computational policy is essentially a service that checks whether one or more conditions are met by a specific version of a data product or one of its components. The service implementing the policy typically receives two inputs: a reference to the data product version or component being verified and (optionally) additional contextual information. The service then returns the verification result, along with optional context explaining the outcome.

There are two main types of computational policies: **static policies** and **runtime policies**. Static policies can be verified even before a data product version is deployed in a runtime environment. They take the descriptor (or a portion of it) as input and perform checks on it. Conversely, **runtime policies** can only be verified after the product version is instantiated. They receive a reference to the running instance and perform checks directly on it.

Static policies are valuable because they can be used to insert quality gates within the deployment pipeline. These gates prevent non-compliant data products from being deployed in the first place. Runtime policies, on the other hand, come into play after successful data product deployment. They validate the product's ongoing compliance at both operational and formal levels.

Static policies are typically verified only once during specific stages of the deployment pipeline. Runtime policies, however, are verified multiple times throughout the product's operational lifespan. They can be executed according to a set schedule (e.g., daily) or triggered by specific events (e.g., data updates). Both types of policies can be either blocking or just informational. Blocking static policies halt the deployment process if non-compliance is detected. Similarly, blocking runtime policies can interrupt operational management tasks (like granting access to an output port) or completely block access to the data product instance until the issue is resolved. Non-blocking policies, on the other hand, simply record and report the non-compliance.

> **Note**
>
> It is possible to implement a service for each policy or to have services capable of executing multiple policies. It is also possible to have services based on proper policy engines. A policy engine allows defining new policies declaratively (policy as code), thus avoiding the need to modify the service code. There are several products that enable defining policies as code. **Open Policy Agent** is one of the most common examples of an open source, full-featured policy engine. **CUE** and **CEL** are instead examples of open source policy checkers that can be used as a foundation for implementing a full-featured policy engine.

Services implementing one or more policies are generally external to the runtime context of a data product instance. However, in some cases, it might be convenient or even necessary to implement a service that verifies one or more policies directly within the runtime context of the product. In these cases, the initiation of policy evaluation can be autonomously managed by the product instance or triggered externally through a control port. The outcomes of the verification are communicated through an observability port. In general, it is advisable to separate sensors from policies. A sensor is an observability component that instruments the product and sends information about its internal state through an observability port. A policy, on the other hand, decides regarding the compliance of a product with a specific predefined rule. A policy may be based on values returned by one or more sensors. Data quality or performance indicators are examples of information about the internal state of the instance detectable through appropriate sensors. The detection of the internal state should always be performed by sensors and not directly by policies. Policies decide, based on the data returned by sensors, whether the predefined rule is met or not. For this reason, cases where it is necessary to execute a policy directly within the runtime context of the product instance should be limited to custom local policies.

A **local policy** is a policy defined directly by the data product team. Local policies complement **global policies** defined by the governance capability but apply only to the specific product on which they are defined. A local policy can extend a global policy by adding additional constraints. A **custom local policy** is a policy implemented directly by the product team because there is no existing platform service capable of verifying the rule associated with the policy. In this case, the policy must be executed within the runtime context of the product instance because it is the only environment where the product team can release components. The platform capability should monitor custom local policies and, when common policies across multiple products emerge, implement a service in the platform capable of verifying them.

The information collected by sensors can be combined to calculate higher-level indicators (e.g., KPIs and KQIs). In this case, it is also advisable to separate indicator calculation logic from policies. Policies use the calculated indicators to make compliance decisions but should not directly calculate them. Indicator calculation should be delegated to specific services implemented by the platform or directly by individual products. Some indicators may be associated with acceptability thresholds. Setting a threshold is useful in defining **Service Level Objectives** (**SLOs**) and **Service Level Agreements** (**SLAs**). SLOs and SLAs are not policies in themselves. Policies use SLOs and SLAs to make compliance decisions. As with status information and indicators, the management of SLOs and SLAs should also be separate from the validation logic implemented by policies.

Policies should be managed as real products. Each policy has a life cycle and therefore should have a version and an owner. A **policy management system** should support the entire life cycle of a policy, not just the validation part. The components responsible for performing validation are called **Policy Decision Points** (**PDPs**). Other components usually present in a policy management system are as follows:

- A **Policy Enforcing Point** (**PEP**) is the component responsible for orchestrating the execution of policies. It invokes the PDPs according to trigger logics, passes the necessary information for validation, aggregates the outcomes, and decides which actions to take accordingly. Generally, access to the PDPs is always mediated and orchestrated by the PEP.

- A **Policy Administration Point** (**PAP**) is the component that manages the policy catalog. It stores available PDPs and all associated metadata such as the policies each of them is capable of validating, as well as the owner, version, and triggering logic. Generally, it allows organizing policies into suites. It provides services or a dedicated UI for searching, creating, modifying, or deleting policies or policy suites from the catalog. In addition to managing policies, the PAP also stores the results of the validations performed.

- A **Policy Information Point** (**PIP**) is the component that allows the PDP to access context information useful for validation (e.g., data from a sensor, the value of an SLA, etc.). It serves to decouple the PDP from external components that produce this information.

Automating policy enforcement is an iterative and incremental process that is best tackled in phases. A wise starting point is static policies, focusing on descriptor file validation during deployment. This initial step ensures that products released by development teams adhere to the governance capability's defined standards, at least on a formal level.

Once this quality gate is established, the next step is to equip data products with observability sensors. The data gathered by these sensors can then be used to define runtime policies. In the next section, we'll delve into how to make data products observable.

Observing data products

By definition, a data product is a module that is designed following the principles of information hiding and encapsulation. Its internal components and operating modes must not be visible to the outside. All interactions with a data product instance are mediated by the APIs of the services exposed through its ports (abstraction). However, to identify and resolve operational problems, it is necessary to be able to access information that allows an external observer to reconstruct the internal state of the data product. A data product must provide such information in a controlled manner through its observability ports. Unlike monitoring, which is a platform capability, observability is a product capability that must therefore be implemented by its development team. All products can be monitored in black box mode from the outside, but only products that provide information about their internal state proactively through observability ports are observable.

All application and infrastructure components of a data product should provide information about their operational state. The data product should also provide information on the state and quality of the data managed (**data observability**), as well as on the use that is made of it. Finally, the data product should provide audit trail information to track who has the grants to access the data and the history of accesses.

State information is collected by means of **sensors** that instrument the component under observation. It is possible to instrument a component in two ways. The first requires modifying the component to actively generate the desired information (white box approach). The second does not require modifying the component but is based on an external agent that monitors it to generate the desired information (black box approach). The first is obviously more flexible and expressive but requires more work and cannot be applied to all types of components that you want to observe.

The information generated by a sensor is called **signals** or **observations**. The main types of signals generally used to communicate the internal state of an object are logs, traces, and metrics. **Logs** are events related to the component observed. **Traces** are sequences of related events. Traces are divided into **spans**. Each span is associated with a specific activity performed by the component observed. Spans are determined by the start and end events of the activity. A trace is the sequence of spans associated with the activities that make up a specific process executed by the observed component. Spans and traces can be deduced from logs when these are properly correlated, generally by means of an attribute present in the log called **correlation ID**. When a process involves multiple distinct components, it is possible to pass the correlation ID from one component to another to correlate logs and spans produced by the various components to generate distributed traces. **Metrics** are numerical indicators that provide synthetic information about the state of the component and vary with it. A log or a trace may contain metrics. Logs and traces can also be used to generate metrics. It is useful to be able to manage metric-type signals within logs and traces but also independently. All signals generated by sensors, regardless of type, are usually accompanied by contextual information. Such information is metadata describing the signal. Typically, the context includes the time when the signal was recorded. Other information, such as the correlation ID, may be included in the context depending on the needs.

The signals generated by sensors that instrument the observed component must be collected to be monitored and analyzed. Sensors must therefore serialize the signals and communicate them to a component that is responsible for aggregating them. It is the responsibility of the governance capability to indicate the serialization format of the signals and the communication protocol to be used to share them with external components. Although it is possible to define custom formats and protocols, it is advisable to consider adopting the **OTLP specification** (`https://opentelemetry.io/docs/specs/otlp/`) defined by the Open Telemetry Project of the Cloud Native Computing Foundation. OTLP is an open and widely adopted specification that describes how to encode signals using **protobuf** and manage their transport using **gRPC**. By adopting OTLP, is possible to leverage a rich ecosystem of tools for defining various types of sensors and managing downstream tools responsible for collecting and analyzing signals.

In an instance of a data product, the signals collected by the sensors instrumenting its components are generally communicated to a component called the **collector**. This component aggregates the signals, optionally transforms or filters them, and then redistributes them through the observability ports of the data product (*Figure 6.4*).

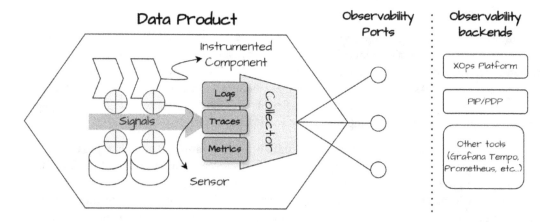

Figure 6.4 – Observability ports

A data product can have multiple observability ports, divided by signal type and/or API type. For example, it's possible to have one port that communicates all signals as a stream of events and another port that stores the signals and allows searches on them. In general, when the protocol defined by Open Telemetry is adopted, it's advisable to have a port that exposes the signals using OTLP to allow platform-level analysis tools that understand that protocol to connect to it.

The number and API of the observability ports that a data product must expose are defined by the governance capability. The governance capability also has the responsibility of defining semantic models to represent the different types of signals that a product can generate. Serialization mode defines the syntax in which the signal is encoded but says nothing about its meaning or that of the context attributes it must have. It's challenging to develop signal analysis tools if every data product generates signals that are syntactically, but not semantically, interoperable. The semantics are defined based on the type of information one wants to communicate with the signal. For example, it is advisable to specify the semantic model to use for communicating an exception or the semantic model to use for communicating the **four golden signals** (i.e., latency, traffic, errors, and saturation) generally used to describe the high-level operational state of a product. The Open Telemetry Project defines various semantic conventions that can be applied in many common contexts (`https://opentelemetry.io/docs/specs/semconv/`).

All components comprising a data product must be properly instrumented and observed. However, the observability of data plays a central role given the very purposes for which data products are built. Generally, this is implemented through quality checks and anomaly detection techniques.

Data quality checks verify various expected conditions (**expectations**) on the data generated at different stages of the enrichment pipeline. For example, they check that field values are not null and adhere to a predetermined format or range. Data quality checks can be grouped according to the quality dimension they verify. Each data quality framework defines different quality dimensions. The most common ones include the following:

- Completeness (is there missing data?)
- Uniqueness (is there duplicate data?)
- Timeliness (is the data up to date?)
- Validity (is the data correct?)
- Accuracy (is the available data the data that's needed?)
- Consistency (is the data consistent?)

For each dimension, specific tests can be defined. There are several tools available for executing data quality checks on data stored in different types of data stores (e.g., **SODA, Great Expectations**). Whenever possible, efforts should be made to define data quality checks declaratively and then use the definition to invoke the underlying tools that execute them. The definition of a check, or even better, the specification of the constraint that generates the check, should be defined within the schema of the data exposed by each output port. This way, consumers know which tests the product performs on the data. The results of the check can then be communicated through various signals via the observability ports (result logs, logs of invalid rows, synthetic quality metrics, etc.).

Anomaly detection, on the other hand, is a black-box technique for monitoring data quality. The data product exposes basic metrics through the observability ports, such as the volume of data contained in a data store, its statistical distribution, the last time it was updated, and how many data points have been loaded. An external system collects these telemetric signals, stores them in time series, and, using ML techniques, evaluates when the value of a signal deviates from the expected one, indicating a likely anomaly in the data and its quality. Anomaly detection does not replace data quality checks but complements them. Data quality checks ensure data meets standards, while anomaly detection identifies unusual patterns in data. Together, they provide a more comprehensive approach to maintaining data integrity.

We have seen how to observe the operational behavior of a data product. Now, let's see how and when it is possible to intervene to control its behavior.

Controlling data products

The development team or the self serve platform may need to intervene in an instance of a data product to perform administrative tasks. These tasks may modify the internal behavior of the product without impacting the functions delivered (e.g., changing the level of detail at which logs are collected and redistributed). It is also possible to execute tasks that modify the behavior of the functions delivered (e.g., increasing the frequency at which data is updated). In this case, however, the new behavior should never contradict what is declared within the descriptor file. For example, it is possible to increase the refresh frequency of the data but not decrease it below the threshold declared in the descriptor file. If it is necessary to decrease the frequency compared to what is declared, then a new version of the product must be released.

All external interventions that can be performed on the instance of the data product are mediated by control ports. Two types of control ports are common to all products: the **config control port** and the **action control port**. The config control port allows passing configuration parameters to the data product instance, which in turn can modify the behavior of its internal components (e.g., logging level). The **task control port** allows triggering an administrative task on the data product. The payload passed to the service implementing this port defines the type of task and the parameters necessary to carry it out successfully. Here are some tasks that instances of data products could support:

- **Change of operational state**: This task allows to enable or stop the loading of new data and access to already loaded data. It might also make sense to define a task to put the data product in a quarantine state in which it loads new data but does not make that data accessible.

- **Forced data deletion**: This task allows for regulatory reasons (e.g., right to be forgotten) to intervene in the data to delete certain records.

- **Assignment of access grants**: This task allows to give or revoke access grants on an output port to a specific consumer.

- **Creation of new quality tests**: This task allows adding new quality checks at runtime if the data product uses a tool that supports a declarative definition of quality checks.

The specification of control ports common to all data products is defined by the governance capability. Configuration parameters and supported actions may vary depending on the products. A dedicated discoverability port should list the configuration parameters and actions supported by a specific instance of a data product along with their descriptions.

In this section, we have seen how to govern data products once deployed. In the next section, we will see how to consume them.

Consuming data products

To make existing data products readily usable, three key areas need to be addressed: discoverability, access request automation, and data composition ease. This section dives deeper into these crucial aspects of data product consumption.

Discovering data products

Scanning the observability ports of all deployed products to find the desired one is neither convenient nor efficient. To facilitate discovery, implementing a data product registry within the self serve platform is highly beneficial. This registry should not only store the main URL of each data product instance but also retain its complete descriptor, indexed appropriately to enable easy search.

Upon storing a data product's descriptor, the registry should be capable of transmitting it, in its entirety or partially, to other metadata management tools. Essentially, the registry should operate as a **data product hub**. This approach allows, for example, sharing a data product's metadata with a **data product catalog** specifically designed to provide a **data marketplace** experience to the users (*Figure 6.5*).

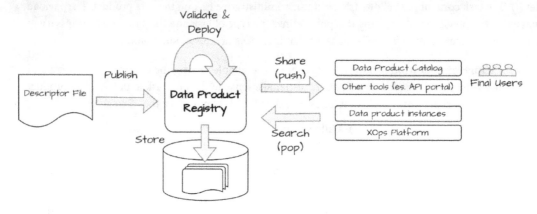

Figure 6.5 – Data product registry

The presence of the registry decouples descriptor publication from downstream tools that utilize it. Multiple tools may require access to information about deployed data products. They can also change over time. This decoupling between data products and downstream tools, facilitated by the registry, enables the flexible evolution of the utilized tool portfolio.

Once the desired product is identified, the consumer needs to access the data it exposes. The next section will delve into this aspect.

Accessing data products

Establishing who can access the data exposed by a data product is the responsibility of the data product owner or the owner of the domain to which the product belongs. Access policies may vary depending on the specific port. Information about who can access a port, and through which authentication protocols, must be documented in the descriptor file. In DPDS, this information is contained within the port API definition. Also, the terms and conditions of use that consumers must adhere to should be included in the descriptor file. In DPDS, this information is contained within the **expectations object**.

Typically, a newly deployed product should not allow access to its output ports to anyone. Access requests must be explicitly made by interested consumers to the platform. The platform, using the information contained in the descriptor, decides whether access can be granted or not based on the role of the consumer and the intended use of data. In cases where the information is insufficient to decide, it may initiate a manual approval workflow involving the product owner or the domain owner. This is usually necessary when the consumer is required to explicitly agree to the terms and conditions of use or when the consumer needs to provide further clarification on the type of data usage intended.

Once access is authorized, permissions to invoke the port API are assigned by calling the task control port, as seen previously. The platform keeps track of granted permissions and manages them over time.

With the necessary permissions secured, the consumer can finally start using the data product instance, combining its data with that exposed by other data product instances. Let's see how in the next section.

Composing data products

Data products can be combined to support multiple business cases. Specifically, two or more data products can be used to support the development of a new data product or a new analytical application, or to support data exploration activities.

In the case of combining multiple data products to build a new data product, the composition occurs through the output ports of the products that provide data to the input ports of the new product. Data reading is automatically managed through the interaction between the output port adapters and the input port adapters connected to each other. Once it is defined within the input port which output port the data should be read from, along with the trigger policy and where the read data should be stored, the data transfer activity can be delegated to the adapters. The team developing the data product only needs to develop the pipelines that transform the data written by the adapters in the ingestion data stores. Communication between output and input adapters can occur directly or be mediated by a centralized transport layer provided by the self serve platform.

In the case where multiple data products are composed to support a new analytical application or data exploration activity, it is useful for the self serve platform to implement a layer of data virtualization. This layer facilitates the collection of data from the different output ports of the relevant data products and consolidates them into analytical views that facilitate consumption by abstracting underlying technical details.

In any case, the platform must track dependencies, both between multiple data products and between data products and any source systems that feed them, as well as any external applications that consume their data. This last type of dependency must be managed with particular care by the platform as it cannot be reconstructed from an analysis of dependencies specified through input ports. The lineage of how data flows between source systems and data products until it reaches analytical applications and final users is of fundamental importance, both to detect problems in the data supply chain and to assess the impacts on downstream consumers due to changes made in an upstream data product. In the next section, we will see how to manage the evolution of data products over time.

Evolving data products

Data products have a life cycle and undergo continuous evolution. This section delves into strategies for managing the evolution of data products while minimizing the impact on their consumers.

Versioning data products

All consumers of a data product access the data and, more generally, the services it offers only through the APIs it exposes through its ports. Therefore, managing the life cycle and evolutions of a data product is not particularly different from managing the life cycle of an API.

Each data product can have multiple versions, but in a given runtime environment, only one instance of the data product per version can exist. The instance with the highest version number present in the production environment is the current version of the data product. The versioning of a data product must use **semantic versioning** (`https://semver.org/`). According to semantic versioning specification, a version must be defined using three numbers: major, minor, and patch version. Whenever the product undergoes a non-backward compatible change in one of its interfaces (e.g., when fields are removed from the schema of exposed data), a new version must be released by incrementing the major version number. When releasing a version with new features that is backward compatible with the previous one, only the minor version is incremented (e.g., when fields are added to the schema of exposed data). Finally, when releasing a version that does not change the product's functionalities but only modifies the internal implementation, such as fixing bugs, only the patch version is incremented. Internal components of a product are also versioned. When the major version of an API changes, the major version of the product must also change. Likewise, when the minor version of an API changes, the minor version of the product must also change. Any other version changes in the APIs or internal components only lead to changes in the patch version of the product.

Deprecating data products

Wherever possible, every modification made to a data product should ensure backward compatibility to avoid impacting existing consumers. If this is not possible, the transition from the previous major version to the new one must be managed following the **deprecation policies** specified within the descriptor file. The most common approach is to notify existing consumers and provide them with a predetermined period of time to migrate. During this period, instances of the old version and the new one coexist in the production environment. All new consumers are associated with the new version, while old ones are gradually transitioned from the old to the new version as they complete the necessary activities to adapt to the new APIs. If the non-backward compatible change is focused on a specific port and it is not convenient to manage two instances of the product in parallel, it is possible to expose both the new and old versions of the port within the same product and gradually manage the migration of consumers from the deprecated version to the new one.

To facilitate consumer migration to the new version of the product, each release must be accompanied by detailed release notes explaining what has changed between versions.

Summary

This chapter comprehensively explored the deployment, governance, and consumption of data product instances.

We delved into the creation of automated deployment pipelines for testing and releasing data product instances across various environments, from test to production. Pipeline automation is the foundation for CI/CD approaches, which are crucial for rapid problem identification and resolution before they become too intricate. By adopting such approaches, development and release activities can be made more efficient and replicable, as we learned in this chapter.

We also examined the governance of data product instances in production environments. Specifically, we focused on methods for collecting and automatically evaluating data product metadata using computational policies. We further explored how to monitor a data product instance through its observability ports, as well as how to perform administrative tasks on it via its control ports.

Finally, we delved into the consumption and evolution of data products over time.

The next chapter will focus on implementing the capabilities that the self serve platform must provide to support data product life cycle automation and utilization.

Further reading

For more information on the topics covered in this chapter, please see the following resources:

* *Software Delivery Guide* – Martin Fowler (2019) `https://www.martinfowler.com/delivery.html`

* *Continuous Integration: Improving Software Quality and Reducing Risk* – P.M. Duvall, S. Matyas, A. Glover (2007) `https://www.amazon.com/Continuous-Integration-Improving-Software-Reducing/dp/0321336380`

* *Continuous Delivery: Reliable Software Releases through Build, Test, and Deployment Automation* – J. Humble, D. Farley (2010) `https://www.amazon.com/Continuous-Delivery-Deployment-Automation-Addison-Wesley/dp/0321601912/`

* *Accelerate: The Science of Lean Software and DevOps: Building and Scaling High Performing Technology Organizations* – N. Forsgren, J. Humble, G. Kim (2018) `https://www.amazon.com/Accelerate-Software-Performing-Technology-Organizations/dp/1942788339`

* *Infrastructure as Code: Dynamic Systems for the Cloud Age* – K. Morris (2021) `https://www.amazon.com/Infrastructure-Code-Dynamic-Systems-Cloud/dp/1098114671/`

- *Policy As Code: Improving Cloud-native Security* – J. Ray (2024) `https://www.amazon.com/Policy-As-Code-Improving-Cloud-native/dp/1098139186`

- *Data Observability for Data Engineering: Proactive strategies for ensuring data accuracy and addressing broken data pipelines* - M. Pinto, Sammy E. Khammal (2023) `https://www.amazon.com/Data-Observability-Engineering-pipelines-actionable/dp/1804616028/`

- *Learning Open Telemetry: Setting Up and Operating a Modern Observability System* – T. Young, A. Parker (2004) `https://www.amazon.com/Learning-OpenTelemetry-Setting-Operating-Observability/dp/1098147189`

- *I test in prod* – C. Majors `https://increment.com/testing/i-test-in-production/`

- *WHOOPS, THE NUMBERS ARE WRONG! SCALING DATA QUALITY @ NETFLIX* – Michelle Winters (2017) `https://www.youtube.com/watch?v=fXHdeBnpXrg`

Get This Book's PDF Version and Exclusive Extras

UNLOCK NOW

Scan the QR code (or go to `packtpub.com/unlock`). Search for this book by name, confirm the edition, and then follow the steps on the page.

Note: Keep your invoice handy. Purchases made directly from Packt don't require one.

7

Automating Data Product Lifecycle Management

In the previous chapter, we saw how to deploy, govern, and use a data product in production. In this chapter, we will analyze how to automate the data product lifecycle through an XOps platform.

We will first analyze the role of the XOps platform within the data ecosystem, delving into its core capabilities and high-level architecture.

We will then see how the XOps platform uses its core capabilities to support all the actors present in the data ecosystem through a curated and unified experience that supports the development, operational management, and consumption of data products.

Finally, we'll explore XOps platform implementation, outlining key principles to guide the critical decision between building (make), purchasing (buy), or adopting a hybrid approach.

This chapter will cover the following main topics:

- Understanding the XOps platform
- Boosting developer experience
- Boosting operational experience
- Boosting consumer experience
- Evaluating make-or-buy options

Understanding the XOps platform

The XOps platform acts as a powerful enabler, streamlining and automating data product development, management, and utilization whenever possible.

In this section, we'll delve into its critical role in fostering a data-as-product management paradigm. We'll explore its core functionalities and how to implement them within a future-proof solution that can effortlessly adapt to your organization's evolving needs.

Mobilizing the data ecosystem

Data per se does not have value. It gains value when it is used to support a business case of organizational interest. However, data generated by operational applications that digitalize the organization's processes must be collected, consolidated, and transformed before it can be used to support relevant business cases. Traditionally, this process of creating value from operational data, through intermediate transformations to final use, has always been seen as a **linear process** (*linear value chain*). Each actor involved (that is, suppliers, producers, and consumers) focuses on a specific part of the process and does not have an end-to-end view of it. The division of responsibilities is orthogonal to the value creation process itself. The actors interact according to upstream/downstream relationships, often not regulated by clear interfaces and precise responsibilities.

We have already seen in *Chapter 1, From Data as a Byproduct to Data as a Product* how this model of value generation from data is systemically dysfunctional and unsustainable over time. The approach of managing data as a product plays a key role in this regard. It not only breaks down solutions into smaller, more manageable parts but also reduces the distance between different actors, fostering the formation of cross-functional teams around relevant business problems. These teams follow the entire process of creating data product-based solutions, sharing goals, responsibilities, and language. In a data-as-a-product management approach, data is a shared asset, and as such, everyone within the organization must be actively involved in its management. Data products define the boundaries within which this collaboration between different actors takes place.

However, most of the tools and solutions available on the market to support data management are designed for the linear value production model (linear value chain). In particular, they focus on optimizing specific tasks related to the integration process. That is, they are designed to specifically support the experience and workflow of data integrators, who bear the majority of the responsibilities related to data management. They are not designed to be accessible to other upstream (data suppliers) and downstream (data consumers) actors. They do not seek to involve them in the active production of value from data in order to facilitate the transition to a more participatory and sustainable operating model.

Conversely, an XOps platform must support all operations related to value production from data, providing integrated functionalities that support and actively involve all actors in data management and utilization (**value network**). It must act as a true platform that mobilizes the entire data ecosystem, helping different actors collaborate directly, reducing the need for intermediation, to solve relevant business problems together using data (see *Figure 7.1*):

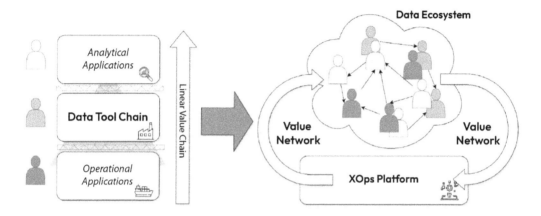

Figure 7.1 – Moving from value chain to value network

The must-have capabilities of an XOps platform can be grouped into the following three categories based on the type of interactions and experiences they aim to support within the ecosystem:

- Developer experience capabilities
- Operation experience capabilities
- Consumer experience capabilities

Products or platforms that provide integrated capabilities to support the developer experience and operation experience are generally called **data developer platforms** or **DataOps platforms**. Products or platforms that provide integrated capabilities to support the consumer experience are generally called **enterprise data marketplaces**. An XOps platform combines these two types of products or platforms into an integrated, though not necessarily singular, solution. This solution is capable of providing a uniform and coherent user experience (**experience continuum**) for every key interaction that connects the actors in the data ecosystem in the distributed value production process (see *Figure 7.2*):

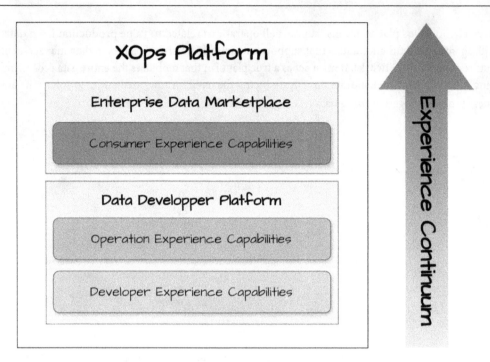

Figure 7.2 – The experience continuum within the XOps platform

A mobilization platform of this kind, which encourages all actors within the organizational ecosystem to work together in data management by fostering long-term relationships to achieve common goals, is a relatively new concept in data management practice. Therefore, there is no standard, widely accepted name for it yet. Possible alternative names for this concept include **data product catalog** or **data experience platform**. Other names may emerge in the future. In this book, we have chosen to use the name XOps platform and will remain consistent with this to avoid confusion.

We have seen so far how an XOps platform fits into the processes of value generation from data. Let's now look in more detail at how it concretely supports these processes.

Understanding platform value engines

An XOps platform groups together products and functionalities useful for implementing the core capabilities necessary to create a shared workspace where all actors can participate and collaborate to jointly produce value from data (value network). The platform must provide two main engines of value creation to enable the establishment and growth of this value network: the transactional engine and the learning engine.

Delving into the platform transaction engine

The **transactional engine** consists of the set of services and experiences the platform offers to the ecosystem, specifically designed to facilitate relationships and value exchanges among the actors within it. It works to reduce development, maintenance, and operational coordination costs (*transaction costs*) to increase engagement and collaboration among the various actors. In essence, the transactional engine focuses on growing the value network within the ecosystem.

Managing data as a product requires certain superstructures that increase local complexity with the promise of reducing global system complexity over time. The reduction of global complexity benefits all actors by lowering costs, increasing release speed, and overall creating a system that is easier to evolve sustainably. However, an initial investment is necessary from all involved actors to achieve these medium-term benefits. To trigger this process of spontaneous collaboration, which encourages each actor to consider medium-term global objectives rather than only short-term local interests, it is crucial to minimize this initial investment. The XOps platform pursues this goal by addressing the transactions that constitute the key value creation processes within the ecosystem. Specifically, it aims to reduce the complexity of these transactions, and consequently their associated costs, through **standardization**, **abstraction**, and **automation**:

- Standardization allows the platform to manage common utility services at a platform level, preventing each data product from having to implement them independently

- Abstraction enables the use of standardized services through declarative interfaces that hide the underlying technological complexity

- Finally, automation speeds up activities that individual actors must perform to achieve their goals

Standardization, abstraction, and automation reduce the **cognitive load** on individual actors, allowing them to focus independently on value-added activities for the organization. The reduction in cognitive load also enables a broader range of actors to actively participate in data management.

The XOps platform, in acting as a transactional engine, not only aims to reduce the costs of key transactions but also must govern interactions among the actors involved in these transactions. It must create a trusted environment where various actors can collaborate securely. It is the platform's responsibility to act as a guarantor of compliance with globally defined governance rules, promoting virtuous behaviors through reward mechanisms and inhibiting inappropriate behaviors through guardrails and quality checks.

Delving into the platform learning engine

The **learning engine** consists of the set of services and experiences the platform offers to the ecosystem specifically designed to help the actors within it learn, improve, and evolve. This is how the platform supports the actors in managing and adapting to the increasing complexity of the artifacts produced and the interactions necessary to produce them. The learning engine focuses on managing the complexity that grows with the expansion of the value network.

As with the transactional engine, the learning engine also focuses on reducing the cognitive load that actors operating within the ecosystem must handle. However, the two engines operate on different types of cognitive load. The transactional engine primarily addresses the types of cognitive load related to the complexity of the specific task (**intrinsic cognitive load**) or the environment in which the task must be performed (**extraneous cognitive load**). Standardization and automation specifically reduce intrinsic cognitive load, while abstraction reduces extraneous cognitive load. The learning engine, however, focuses on **germane cognitive load**, which is the effort required to understand a problem, make appropriate decisions, and execute the task consequently to produce effective solutions. This is the highest value-added cognitive activity because it is directly related to problems and how to generate solutions rather than the tools used or the context in which one operates. In carrying out this activity, the actors involved construct a model of the problem space, combining new information with existing knowledge to build the desired solutions. It is through this activity and the interactions among the actors involved that true learning occurs.

The XOps platform must facilitate learning processes at all levels, from individual learning to group learning to organizational learning. To this end, it must support the processes of knowledge creation, collection, and dissemination. There are two types of knowledge that the XOps platform must manage to support the data ecosystem: **business knowledge** and **contextual knowledge**.

Business knowledge pertains to the domain model in which the organization operates, including the main entities that compose it, how they are related, and through which core processes. Collecting and redistributing this type of knowledge at the organizational level is essential for establishing a shared and uniform basic understanding of how the organization operates. This common foundation is crucial for creating derived models and building solutions to specific problems. Sharing business knowledge accelerates the solution modeling process, facilitates interaction between different actors, and enables the construction of a modular system composed of interoperable parts at both the syntactic/technological and semantic levels (**composable architecture**).

Contextual knowledge, on the other hand, pertains to the operational modalities of the data management function, including its objectives, how they align strategically with the organization's goals, and how they align practically with the portfolio of planned activities. Collecting and redistributing this type of knowledge is essential to ensure that various parts of the ecosystem can work independently and in a distributed manner while maintaining alignment with the strategic objectives defined by the organization they are part of. Similarly, creating visibility on results obtained from the various activities and interactions within the ecosystem allows the rest of the organization to intervene through the control function (**System 3**) to adjust priorities or objectives or through the coordination function (**System 2**) to revise the execution plan of the data management function (**System 1**).

Both types of knowledge are formalized or created through the interactions and value exchanges that occur among the actors in the ecosystem. The learning engine's task is twofold: to facilitate the collection of knowledge created in each interaction and to enhance interactions by leveraging previously collected knowledge. For knowledge to be collected, redistributed, and reused, it must be modeled and described, preferably in a formal and unambiguous manner using **ontologies**. The XOps platform, through the learning engine, must facilitate the construction and evolution of ontologies, making their use as pervasive as possible in every interaction and user experience. We will delve deeper into ontologies and how to use them to manage knowledge in *Chapter 10, Distributed Data Modeling.*

Leveraging platform value engines

The XOps platform, through its value creation engines, works to promote the adoption of the new paradigm centered on managing data as a product within the ecosystem. The greater the adoption, the more significant the positive network effects. Among these effects, the most important is the increased ability to reuse existing products and knowledge to solve new business problems more quickly. To encourage the adoption of the new paradigm, the XOps platform's transactional engine and learning engine work synergistically, promoting strategic alignment (**purpose**), enabling everyone to operate as autonomously as possible (**autonomy**), and supporting continuous learning (**mastery**). Purpose, autonomy, and mastery are the three main drivers the platform uses to motivate actors to participate in data management, undertake new activities, and assume new responsibilities to produce greater overall value for the organization (see *Figure 7.3*):

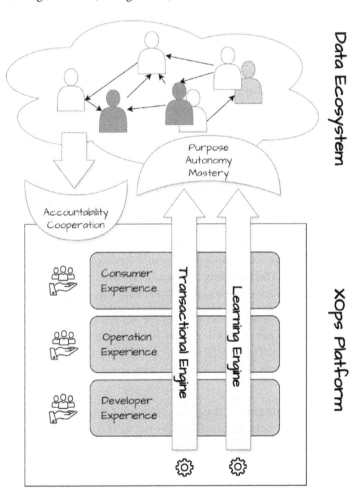

Figure 7.3 – Platform value creation engines

So far, we have seen how the platform operates and through which engines it creates value for the ecosystem it aims to mobilize. It is now time to delve into its architecture.

Exploring platform architecture

In *Chapter 3, Data Product-Centered Architectures*, we saw how the functionalities offered by the XOps platform can be conceptually divided into two planes: the **control plane** and the **utility plane**. The utility plane contains the adapters used by the control plane functionalities to access external services. An adapter enables access to an external service by providing a service based on a standard interface (*first-order services*). The control plane functionalities orchestrate external services using the standard interfaces implemented by the adapters (*second-order services*). This allows the control plane to focus on offering higher-level services to users, ensuring a unified experience centered on the concept of data products.

Each adapter is an instance of a specific capability of the utility plane. All instances of a utility plane capability share the same API. The main capabilities that the utility plane should have are the following:

- **Command execution**: Through adapters that implement this capability's API (`executor-api`), the platform can execute commands on an external application or service and monitor their execution. This capability can be implemented by specific adapters to interact with **continuous integration/continuous deployment** (**CI/CD**) applications to start deployment pipelines, permission management applications to grant access to a resource, task management applications to open a ticket, and so on.

- **Event notification**: Through adapters that implement this capability's API (`observer-api`), the platform can transfer information to an external application or service whenever certain events of interest occur (for example, the publication of a new data product). This capability can be implemented by specific adapters – for example, to communicate metadata associated with a data product to external data catalogs when the product is released.

- **Policies validation**: Through adapters that implement this capability's API (`validator-api`), the platform can verify the compliance of a resource managed by the XOps platform with a policy or suite of policies. This capability can be implemented by specific adapters to validate a resource using external policy engines such as **Open Policy Agent** (**OPA**) or custom functions.

- **Signals aggregation**: Through adapters that implement this capability's API (`collector-api`), the platform can collect signals (logs, metrics, and traces) generated by external sensors and make them queryable. This capability can be implemented by specific adapters to read telemetry or quality data from an external observability application related to a specific instance of a data product.

- **Configuration management**: Through adapters that implement this capability's API (`configurator-api`), the platform can create, modify, and read configurations from an external system. This capability can be implemented by specific adapters – for example, to interact with an external vault to read secrets necessary to access other external services.

Multiple adapters can be associated with the same capability by implementing the corresponding standard API (see *Figure 7.4*):

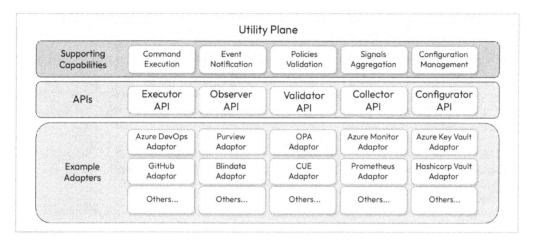

Figure 7.4 – Utility plane APIs and adapters

This allows the control plane to operate in hybrid environments. An API exposed by the utility plane can, for example, be implemented by different specific adapters that act as mediators toward different cloud provider services. The functionalities of the control plane that use this API can dynamically choose which adapter to use based on the context or user preferences. Furthermore, since the interaction between control plane functionalities and adapters that instantiate utility plane capabilities occurs via standard APIs, it is possible to change adapters without impacting the functionalities of the control plane and, consequently, their users. This means that it is not necessary to directly intervene on existing data products every time an application or service of the underlying technology platform changes. Simply creating a new adapter is sufficient.

The platform capabilities implemented through the control plane functions can be grouped, as we saw in the previous section, based on the type of user experience they support. At the developer experience level, the main capabilities that the platform should have are the following:

- **Policy development**: Through this capability, the platform supports the development and evolution of computational policies used to automatically verify compliance with governance global rules of data products

- **Ontology development**: Through this capability, the platform supports the development and evolution of ontologies at both corporate and individual domain levels, facilitating conceptual modeling processes in a distributed environment

- **Data product development**: Through this capability, the platform supports the development and evolution of data products by providing product teams with reusable blueprints, modules, building blocks, and sidecars

At the level of operation experience, the main capabilities that the platform should have are the following:

- **Policies enforcement**: Through this capability, the platform verifies the defined computational policies. Policies can be verified both before the product release (static policies) and during its operational life (runtime policies).

- **Data product metadata management**: Through this capability, the platform stores, indexes, and makes available for search all metadata associated with a data product. The main metadata of a data product is contained within its descriptor file provided to the platform during publication. The platform can then generate additional metadata associated with a data product during its operational life.

- **Data product lifecycle management**: Through this capability, the platform manages and automates where possible the lifecycle of a data product, handling its deployment and all subsequent state and/or runtime environment transitions.

- **Data product monitoring**: Through this capability, the platform monitors the operational status of released data product instances, aggregating and enabling analysis of all signals generated by sensors and communicated via observability ports.

- **Data notification management**: Through this capability, the platform collects events occurring within the platform itself and communicates them to interested observers.

At the level of consumer experience, the main capabilities that the platform should have are the following:

- **Data product portfolio management**: Through this capability, the platform promotes strategic alignment among all actors in the ecosystem by providing a complete view of the product portfolio's status and showing connections between products, use cases, and organizational objectives.

- **Data product searching**: Through this capability, the platform allows users to search for a data product within the catalog of released products and analyze its structure to understand the relevance of the results based on specific user needs.

- **Data product access management**: Through this capability, the platform allows users to request and obtain access permissions to data exposed by a data product. It is also through this capability that the platform tracks granted permissions and manages their lifecycle.

- **Data product composition**: Through this capability, the platform facilitates the no-code composition of existing products to generate new data products or create data views to support analytical applications.

- **Collaboration management**: Through this capability, the platform facilitates collaboration among actors in the ecosystem, such as allowing assignment of responsibilities, team creation, task management, or simply exchanging notes and messages.

The following diagram shows the conceptual architecture of the XOps platform with the respective core capabilities described thus far:

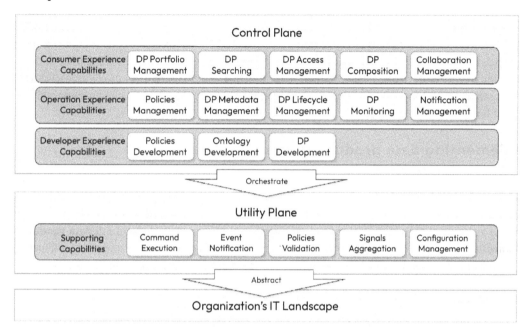

Figure 7.5 – XOps platform capabilities map

The platform capabilities must be implemented incrementally, focusing on the core interactions among the actors that need to be supported, to provide a unified and consistent experience.

"Platforms are designed one interaction at a time. Thus, the design of every platform should start with the design of the core interaction that it enables between producers and consumers. The core interaction is the single most important form of activity that takes place on a platform—the exchange of value that attracts most users to the platform in the first place." (Geoffrey G. Parker et al., Platform Revolution)

In the following sections, we will identify some of the core interactions that characterize the experiences the XOps platform aims to support. For each of these interactions, we will delve into how the capabilities described so far can support them. Let's start from the next section by analyzing key interactions that need to be supported at the developer experience level.

Boosting developer experience

Each data product to be released must provide, through a descriptor document, all necessary information both to assess its compliance with defined governance policies and to manage and operate it correctly within the runtime environment. There are typically no restrictions on its internal structure, which is hidden through the abstraction provided by its exposed interfaces (*information hiding*). The team developing a product can, therefore, autonomously choose the technologies and internal architecture. However, implementing this internal infrastructure from scratch for each product generates a cognitive load on the development team, requiring both greater specialized technical skills and longer development times. To improve the development experience, the XOps platform must provide reusable components that can be combined to build new data products, reducing technological complexity and development times. The platform must also support the governance capability in defining policies that ensure the quality of developed data products and modeling teams that define ontologies useful for ensuring that data products are semantically interoperable and, therefore, cross-domain composable. Let's see how in this section.

Implementing data product building blocks

A **building block** is a reusable component in the development of various data products. It can be an infrastructure component (for example, a database), an application component (for example, a set of data transformation primitives), an interface (for example, the standard API of an observability port), or a composition of these three types of components. For simplicity, we'll call **data product modules** the building blocks defined by composing simpler building blocks. In this section, we'll focus on **atomic building blocks**, which are not composed of simpler building blocks. In the next section, however, we'll see how to compose them to create more complex modules, eventually leading to the creation of complete blueprints for data products.

Atomic building blocks can provide components useful for implementing a specific type of data product or generic components useful to all data products. Building blocks that provide cross-concern functionalities (for example, data input and output management, monitoring, security, and so on), potentially useful to all data products, are generally called **data product sidecars**. We'll delve into sidecars in the upcoming sections. Here, we'll begin to delve into atomic building blocks that provide specific components useful to certain types of data products but not necessarily to all. For convenience, we'll simply use the term *building block* to refer to them. We'll use the terms *module*, *blueprint*, or *sidecar* to refer to the other types of building blocks described so far and represented in the following diagram:

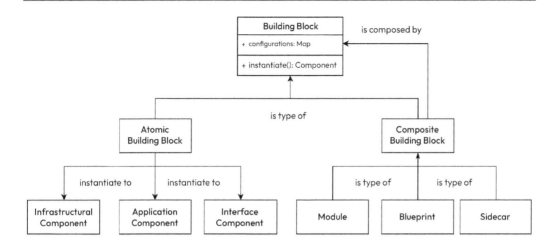

Figure 7.6 – Building blocks' taxonomy

A building block is a standard component provided by the platform that can be instantiated and used within a data product. Using an existing building block allows the product team to avoid developing the associated components themselves. At the same time, it centralizes management, facilitating maintenance and evolution in accordance with governance policies and platform best practices. Regardless of the type of associated component (infrastructure, application, or interface), each building block must have a unique name and version through which it can be unambiguously referenced. Additionally, each building block must implement a standard interface that can be used to instantiate it. This interface should be declaratively defined, hiding the internal model used by the building block to instantiate the component. In other words, the interface should receive configurations as input and instantiate the component accordingly. Building blocks are platform components, and as such, they are domain-agnostic, meaning they do not directly contain business logic. Some building blocks may allow business logic injection through the configurations used during instantiation.

During the creation of the data product, the development team must be able to specify the configuration of building blocks they use directly within the descriptor file. These pieces of information are then used by the XOps platform during deployment to properly instantiate the building block.

In the **Data Product Descriptor Specification (DPDS)**, for example, each component can be defined inline or through a reference object that points to an external definition. The referenced component can be defined by the product development team in a separate repository, or it can refer to a building block defined at the platform level. The following code snippet shows an example of how a descriptor can reference in DPDS an external building block, passing configuration parameters that the platform will then use during instantiation:

```
infrastructuralComponents: [
    {
        "description": "The topic used to store e-commerce sales events",
        "mediaType": "application/odmp",
        "$ref": "https://odm-platform.lux.com/building-blocks/kafka-
            topic-1.0",
        "variables": {
            "name": {
                "default": "onlineSales",
                "description": "the name of the component"
            },
            "version": {
                "default": "1.2.0",
                "description": "the version of the component"
            },
            "topicName": {
                "default": "online-sales",
                "description": "the name of the new topic"
            },
            "partitions": {
                "default": "3",
                "description": "the topic partition number"
            },
        },
    },
...]
```

The XOps platform should be able to manage a catalog of defined building blocks. During the publication of the data product, the XOps platform should replace all references to building blocks in the descriptor with their complete definition, overriding variables with the parameters specified by the development team. During deployment, the XOps platform should be able to instantiate all building blocks referenced by the data product, passing the specified configuration parameters to the deployment pipeline or default values for those not directly specified in the descriptor (see *Figure 7.7*):

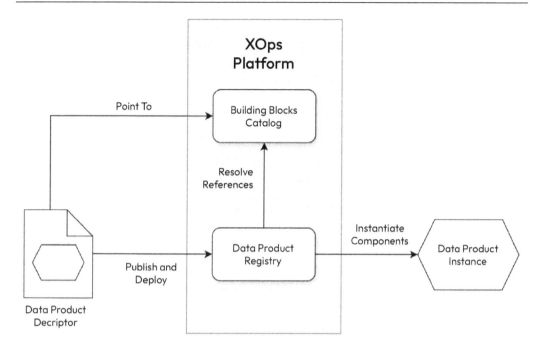

Figure 7.7 – Building blocks' catalog

Once the deployment of the building block is completed, the XOps platform should finalize by populating variables whose values can only be determined at runtime (for example, Kafka topic ID). The discoverability endpoint of the data product instance that exposes the descriptor should always return a version with complete descriptions of the components instantiated by a building block, together with a reference to the building block and the variables used to instantiate it into a component instance.

Each component within a data product descriptor may depend on one or more other components defined within the data product. For example, in DPDS, dependencies between components of the same type (for example, infrastructural components) are expressed through the dependsOn attribute, while dependencies between application components and infrastructural components or between interface components and application components are expressed through the consumesFrom and providesTo attributes. For instance, an application component might need another application component to handle configurations (for example, dependsOn: configSidecar) and two infrastructural components, respectively, to execute it (for example, consumesFrom: k8s) and to save its outputs (for example, producesTo: postgres). Building blocks used to define reusable components across multiple products may also have dependencies. It is the responsibility of the development team using a building block to ensure that all its dependencies are satisfied. They must define the necessary components for the building block and pass their IDs as configuration variables during referencing. Of course, these components may themselves be defined using other building blocks.

So far, we have seen how to *use building blocks by reference*. That is, the development team does not have the freedom to directly modify the building block except through the configuration parameters it exposes. This mode of usage is preferable as it ensures the reuse of standard components, certified at the platform level. However, in some cases, it may make sense to allow the *use of a building block by copy*. In this mode, the building block is cloned into the product team's development environment, allowing them to modify and customize it. Thus, the reference to the original building block is lost. However, this allows the product team to quickly build a customized version of a building block if the version provided by the platform does not fully meet their needs without having to reimplement it from scratch. The use of building blocks by copy should be monitored and avoided whenever possible, favoring the evolution of the building block offered by the platform to meet new needs instead.

The platform team should manage building blocks as true products whose users are the development teams. Whenever possible, components should be commonly shared by creating a new building block. Similarly, existing building blocks should evolve over time to best meet the needs of their users.

Similar to data products, building blocks must also have an associated descriptor file. This descriptor document should describe the building block (for example, name, version, owner, and so on), list its possible configuration parameters, define its deployment pipeline and how to invoke it, specify its dependencies, and indicate whether it can be used only by reference or also by copy. The descriptor document of a building block should contain all the information useful to both product teams and the XOps platform for its utilization. The catalog of building blocks that the XOps platform should provide is responsible for listing published building block descriptors and making them easily searchable and referenceable.

In the next section, we will see how to compose atomic building blocks to create modules and blueprints.

Defining data product modules and blueprints

A data product module is defined by composing atomic building blocks. Defining a module facilitates development teams when a set of building blocks of different types are designed to work together to implement a vertical functionality offered by the data product. For example, the standardized management of an observability port based on the *OpenTelemetry* protocol could be realized through a module composed of several atomic building blocks. Specifically, this module could contain a building block containing the port's API (interface component), one containing the application that contains the services implementing the API (application component), and finally, one to manage the runtime environment of the application (infrastructural component). Instead of referencing the three building blocks separately and managing dependencies, the development team of a data product can directly reference the module by passing the necessary configuration variables, generally the composition of the variables exposed by the building blocks that compose it. The XOps platform manages the instantiation workflow of a module by executing the instantiation of the building blocks that compose it in the correct sequence.

A blueprint is a module that contains all components necessary to define completely a data product. It is used to define common types of products that can then be extended to add, as needed, the necessary business logic through the configurations of the individual building blocks that constitute it or through the addition of specific components. Depending on the underlying technological architecture and the types of data transformations that need to be implemented, it is useful to define specific product blueprints. In the case of Lux, for example, it might be useful to define a standard blueprint for data products that need to process data in streaming and one for products that need to process data in batch mode. The blueprint for streaming data management should include building blocks for managing data ingestion and consumption in real time as sequences of events and building blocks to instantiate the underlying infrastructure and applications needed to transform data in real time. Similarly, the blueprint for batch data management should include building blocks for managing delta data ingestion and its exposure through published views on a data virtualization layer and building blocks for managing necessary batch transformations.

Even for modules and blueprints, it is always preferable to use them by reference rather than by copy. However, copy usage should not be prevented but carefully monitored. As with building blocks, modules and blueprints must be managed by the platform team as products, described through appropriate descriptor documents, and published on a catalog offered by the XOps platform.

Now, let's explore operational needs common to all data products that can be useful to implement through appropriate building blocks or modules called sidecars.

Leveraging sidecars to manage cross-cutting concerns

A sidecar is a building block or module that implements a function useful to multiple data products, though not necessarily all. It is essentially a piece of platform whose execution does not occur centrally but within the runtime context of the data product itself. The use of some sidecars may be made mandatory by governance policies depending on the type of data product.

We can group the types of sidecars into two main categories: **adapter sidecars** and **utility sidecars**. An adapter sidecar decouples the internal components of the data product from the external environment. Conceptually, it coincides with the adapters defined within *hexagonal architectures* to which pure data products are inspired. In practice, adapter sidecars are used to implement the ports of a data product. Let's see for each port what functionalities the adapter sidecars can provide:

- An **input port adapter sidecar** is responsible for acquiring data of interest from other data products through their output ports or from external applications and making it available for processing in one or more internal ingestion datastores. In general, it should accept as configuration the identifier of the system from which to read, the ingestion mode (for example, delta or full), any filters to be applied to the ingested data, and the ingestion frequency. Optionally, it could provide load balancing, decryption, health checks on the read data, fault injection, and ingestion monitoring functionalities.

- An **output port adapter sidecar** is responsible for exposing data present in the consumption datastores to external consumption through services that allow controlled access. In general, it should accept as configuration the portion of data to be exposed, who can request access, and any limits on the amount and frequency of data read. Optionally, it could provide load balancing, encryption, health checks on the exposed data, and access monitoring functionalities.

- A **control port adapter sidecar** standardizes the management of data product configurations and the exposure of actions that can be invoked on it at runtime by the development team or the platform team. It generally exposes two services. The configuration management service allows reading and modifying configurations that can then be used by other product components to modify their runtime behavior (for example, log level, reading frequency, and so on). The action execution service allows for specifying the action to be performed and the parameters to be used in execution (for example, enable or disable the product, delete data, grant access on a port, and so on). Each component of a data product should specify through a dedicated discoverability port which configuration parameters and actions it supports.

- An **observability port adapter sidecar** standardizes methods of collecting signals generated by the product's components and their exposure to external consumers. It can operate as a simple collector by aggregating and redistributing the generated signals, or it can also archive them and allow external consumers to query them.

- A **discoverability port adapter sidecar** standardizes methods for accessing the data product's descriptor and other static information useful for its use, such as configuration parameters and supported actions.

It is possible to have a single configurable sidecar per port type or to have multiple specific sidecars per port type. For example, you can have a single input port adapter for all supported API and protocol types or a specific adapter for each type of service/source system. In general, it is advisable to unify the interfaces of ports as much as possible but to have multiple implementations of specific sidecars according to functional needs. Unifying interfaces facilitates the use of the product by external agents. Through the specialization of implementations, however, modularity and composability of building blocks are favored, allowing them to be developed and tested independently by the platform team.

Utility adapters, on the other hand, implement utility functions that, for performance and cost reasons, do not make sense to be executed centrally by the platform. The implemented utility sidecars depend heavily on the context and needs of the organization.

An example of a commonly implemented utility sidecar is one that offers primitives for executing quality checks. Typically, this sidecar transforms the constraints defined in the data product descriptor document into calls to functions offered directly by the sidecar or indirectly through external frameworks (for example, *Great Expectations* or *Soda*) to execute data checks. The sidecar is then responsible for communicating the results of the executed checks (*quality indicators*) to the observability port adapter sidecar to make them accessible externally. Optionally, it can integrate with the control port adapter sidecar to read configurations of interest (for example, test execution frequency). Always thanks to integration with the control port adapter sidecar, it is possible to disable or enable some checks or create new ones at runtime. Quality checks could also be performed centrally through a platform service. However, this would require data movement. Packaging such a service into a utility sidecar and distributing the checks' execution directly to the products is more convenient.

Another example of a utility sidecar useful in different contexts is the sidecar that handles data anonymization. Generally, this sidecar integrates with output port adapter sidecars to generate anonymized versions of the data to be exposed.

A final example of a utility sidecar is one that resolves the identity of business entities read from different input ports associated with systems that do not share the same master data. The identity resolution heuristics and the subsequent strategy to assign a unique identifier to each entity must obviously be shared by all source-aligned data products that deal with the same entities. It is useful, therefore, to uniquely define such an identity resolution service. However, it is more convenient to include this logic in a sidecar and perform identity resolution locally to individual data products rather than centralizing it at the platform level for scalability and performance reasons.

Sidecars, as with all other types of building blocks seen so far, facilitate the development of data products. They facilitate the development experience for product teams. However, the platform should not only support product development teams in developing data products but also support modeling teams in developing ontologies and the governance team in developing computational policies. Let's see how in the next section.

Supporting computational policy and ontology development

To support the development activities across the entire data ecosystem, the platform must also provide useful functionalities for governance and modeling teams.

The governance team defines global policies that must be adhered to by data products to ensure their quality and interoperability, at least at a technical level. The platform should allow defining governance **policy as code (PaC)** to subsequently verify it automatically (computational policy). Computational policies should be managed as a product. The platform must support the governance team not only in defining computational policies but also in managing their entire lifecycle. It should be possible to define policies declaratively, document them, and specify their scope and verification methods.

Modeling teams define ontologies, conceptual models that describe the core data managed by a subdomain (subdomain ontology) or the entire organization (enterprise ontology). Ontologies should also be managed as products and used pervasively within all interactions occurring in the ecosystem. The platform should simplify their development and evolution throughout their lifecycle. This means it should be possible to define new ontologies, import public ontologies, extend existing ontologies by introducing new concepts or relationships, link ontologies together to build a federated conceptual model, and finally associate the defined concepts with data assets exposed by data products (semantic linking).

We've seen in this section how the platform can support the development of data products, computational policies, and ontologies. Now, let's see in the next section how the XOps platform can operationally support these three types of products throughout their lifecycle.

Boosting operational experience

The operational experience revolves around two main types of interactions: the release of data products developed in the runtime environment and their subsequent operations control. In this section, we will see how the XOps platform can support these two key interactions through its core capabilities.

Orchestrating the data product deployment pipeline

Every data product must have an associated deployment pipeline to manage its release in the runtime environment. This pipeline is generally constructed by composing simpler pipelines associated with the release of individual components or groups of components. The reference to the pipeline to be used or to its sub-pipelines must be explicitly specified in the descriptor document.

The following code snippet shows how pipelines to promote the data product through the different stages of its lifecycle are referenced in DPDS:

```
"lifecycleInfo": {
     "test": [...],
     "prod": [...],
     "deprecated": [...]
  }
```

In a declarative manner, the example defines three stages in the product lifecycle: `test`, `prod`, and `deprecated`. For each stage, it specifies a sequence of tasks that must be executed to transition the product to that stage. Each task specifies the name of the executor adapter to be invoked to execute it (`service`); in this case, the adapter for Azure DevOps, the actions to be executed (`template`), and a set of configuration parameters to be used during the task execution (`configurations`). The following code snippet shows how the task that promotes the data product to the test stage is defined:

```
{
"test": [{
       "service": {
          "$href": "{azure-devops}"
       },
       "template": {
          "name": "inventoryDpPipeline",
          "version": "1.0.0",
          "specification": "azure-devops",
          "specificationVersion": "1.0.0",
```

```
            "definition": {
              "organization": "lux",
              "project": "inventory",
              "pipelineId": "3",
              "branch": "main"
            }
          },
          "configurations": {
            "stagesToSkip": ["Deploy"]
          }
        }],
  "prod": [...],
  "prod": [...]
  }
```

The XOps platform must be able to resolve references to the deployment pipeline and orchestrate its execution, possibly delegating some tasks to underlying CI/CD tools through appropriate adapters (*data product lifecycle management capability*). It is advisable to centralize the orchestration of deployment to track all released data products and ensure their compliance before and after deployment.

To deploy a data product instance through the XOps platform, it is first necessary to publish its descriptor document. In this phase, the XOps platform verifies the correctness of the descriptor document and its compliance with static governance policies (*policies enforcement capability*). If the descriptor document is valid, it is stored within the product registry (*data product metadata management*) and can be deployed in one of the available runtime environments. Additional checks can be performed before executing the deployment, for example, to ensure that a product can actually be deployed in the specified environment. Once the deployment is completed, all checks related to runtime governance policies verifiable only on the deployed product instance can be executed. If all checks pass successfully, the product has been successfully deployed; otherwise, a rollback must be performed. When the XOps platform successfully deploys a data product in a runtime environment, it notifies all external applications involved, passing them the descriptor document or parts of it, along with information regarding the deployment just executed (*data notification management capability*).

For all released data product instances, the platform must finally ensure monitoring throughout their operational life until decommissioning. We will see how in the next section.

Controlling data product operations

Once an instance of a data product is created, the platform must provide the appropriate interfaces to access its control and observability ports.

Access to control ports is functional for easily passing configuration parameters to the product to modify its operational behavior without releasing a new version. Through the platform, it must also be possible to invoke control ports when necessary to perform actions on the data product instance.

Access to observability ports, on the other hand, is used to collect signals generated by data products and provide a view of their internal operational state (*data product monitoring capability*). The historical series of telemetry data can be exposed to potential consumers. Additionally, the platform can use the values of indicators exposed by observability ports (**service level indicators (SLIs)** and **data quality indicators (DLIs)** to calculate **service level objectives (SLOs)** and **service level agreements (SLAs)** declared by the data product in its descriptor document. If an SLO or SLA is violated, the platform must manage the notification of the event to interested external applications and agents (*notification management capability*). It must also provide the necessary tools to track and support **incident management (IM)** activities (*collaboration management capability*).

In the next section, we will finish our journey through the experiences that the XOps platform must support by delving into consumer experience.

Boosting consumer experience

The consumer experience revolves around two main types of interaction: the use and the composition of data products. In this section, we will see how the XOps platform can support these two key interactions through its core capabilities.

Managing the data product marketplace

The XOps platform supports the use of released data products through marketplace functionality. It must provide an easily navigable catalog containing all released data products, displaying the relationships between various data products, the teams that develop them, the business cases they support, the physical assets exposed, and the concepts in the defined ontologies that they reference (*data product portfolio management capability*). Furthermore, the XOps platform should enable grouping and filtering of data products according to different criteria and support searches within the catalog (*data product searching capability*). The search functionalities should make use of the encoded knowledge within the referenced ontologies, combining traditional information retrieval techniques with semantic search.

Once a potentially interesting product is identified, the XOps platform should provide a detailed product page containing all necessary information for a potential consumer to understand which data is exposed and evaluate its utility based on their needs (*data product metadata management capability*).

Finally, the XOps platform must allow consumers to request access to the product or to a specific port (*data product access management capability*). The platform should manage the approval workflow for the request, and if the outcome is positive, grant access (data product access management). Optionally, the platform can provide usage monitoring and billing functionalities for certain products according to the pricing models specified in the descriptor document. Additionally, users should be able to add comments to a product and interact with the development team by asking questions or reporting issues (*collaboration management capability*).

The marketplace functionalities support data product usage. However, data products are not only meant to be used separately. On the contrary, they provide more value when combined to generate new data products or create views to support new analytical applications. In the next section, we will see how the XOps platform can assist users in composing products available in the catalog.

Supporting data product composition

The platform should facilitate the composition of existing data products both to create new derived data products and to create data views on which to develop new analytical applications. Composition between existing data products should be possible in a low-code or no-code manner. In other words, it should be achievable by all actors within the ecosystem, especially business users without technical knowledge. When multiple products are combined to generate a new data product shared with the rest of the ecosystem, the platform should assist users in providing all the necessary information for publishing the data product in the catalog so that it is compliant with all global policies defined by the governance capability (*data product composition capability*).

So far, we have described the capabilities that an XOps platform should have and how they should be orchestrated to support the production of value within the data ecosystem. In the next section, which concludes the chapter, we will see how to proceed to implement an XOps platform.

Evaluating make-or-buy options

In this section, we will see which strategy to follow to adopt an XOps platform that can assist in transitioning to a paradigm based on managing data as a product. Specifically, we will explore when is the right time to start evaluating such a platform, how to choose between make or buy, and finally, which elements to consider to capitalize on technology investments already made while maintaining a certain level of flexibility to adapt to future technological innovations.

Deciding when to make the call

In every paradigm shift, technology can assist in the transition. In this chapter, we have seen how an XOps platform is essential for removing friction in data product development, mobilizing the entire ecosystem around data products, and promoting value production by effectively and efficiently combining them. However, technology supports the transformation process but cannot guide it. Therefore, it doesn't make sense to choose or implement an XOps platform before starting the journey. It is useful to begin the transformation on a small scale, experimenting with operational models and developing the first lighthouse data products. In this mobilization phase, the presence of the fully featured XOps platform is not necessary. It may still make sense to consider developing a minimum viable platform that can be further evolved in later phases. It becomes necessary to invest more in the platform when transitioning from the initial mobilization phase to scaling the initiative, seeking to spread the adoption of the paradigm across a broader ecosystem within the organization. At this point, the new operational model has been tested on a small scale, making it easier to outline the most useful capabilities that the platform should provide to support while scaling up. In short, it is possible to make a more mature choice by selecting the technology that best supports the operational model adopted rather than delegating the definition of the operational model to the technology.

Solving the dilemma between make or buy

The market has yet to offer complete solutions to manage data as products such as the one described here. However, this doesn't necessarily mean that an XOps platform needs to be implemented internally. It's possible to select different market solutions for each functional area and then integrate them to cover the core capabilities described here, providing a uniform user experience. For example, it is possible to choose a data developer platform and integrate it with a data catalog and/or a data marketplace. The important thing is that the selected tools allow for managing the concept of data products natively or through customization.

In general, it's not necessary to cover all capabilities right away. It is possible to proceed incrementally based on needs, starting with a minimal platform (*thinnest viable platform*). This platform can then be extended over time by rolling out new capabilities as needed, following a make-or-buy logic (*harvested platform*). The XOps platform as a whole should be managed like a product and evolve over time to best meet the real needs of its users within the data ecosystem. The criterion for prioritizing platform development activities and, more generally, for evaluating its success is user adoption.

It's normal in the XOps platform implementation to start with the part that supports the development experience, particularly from the data product registry, deployment management, and computational policies' enforcement, in order to define necessary quality gates to ensure the quality of released data products. However, the other types of experiences described here, which are equally important for creating a value network that actively involves all actors in the ecosystem, should not be overlooked. For this reason, platform capability development should occur in parallel across all types of experience we want to support.

The XOps platform must obviously fit into an ecosystem of pre-existing and continuously evolving technologies. In the next section, we'll see what characteristics the XOps platform should have to integrate seamlessly with existing external technologies or those that may potentially be adopted in the future.

Open Data Mesh Platform

Open Data Mesh Platform (`https://platform.opendatamesh.org/`) is an open source, multi-plane, and multi-module platform designed to automate the lifecycle of a data product. It uses DPDS (`https://dpds.opendatamesh.org/`) as the specification for representing a data product and offers functionalities to support development, manage releases, and validate data products. This platform exemplifies the architecture presented in this chapter and can be adopted as a foundational platform that can then be extended with new functionalities or integrated with external products. Opting for an open source platform strikes a good balance between building and buying.

Future-proofing your investments

To capitalize on past technological investments and facilitate the adoption of new innovative technologies in the future, the XOps platform must have several key characteristics (architectural principles).

First and foremost, it must provide a curated user experience. Even when the XOps platform is composed of multiple tools, it should never be just a simple bundle of these tools. The primary objective of the XOps platform is to orchestrate existing tools to create the desired interaction experience rather than attempting to replace them by directly implementing the necessary functionalities.

It is also important that integrations between the different modules and products that make up the platform occur through standard APIs that ensure decoupling between the involved components. This way, it is relatively easy to change a module or product to adopt a new, more innovative, or more cost-effective technology when the opportunity arises in the future.

Finally, it must define common specifications to declaratively represent the main managed resources – namely, data products, building blocks, ontologies, and computational policies. The representation of these resources through common specifications is the most important information the platform must manage; they are its master data. How this master data is then translated to be managed by the individual products that make up the platform is just a derived technical problem. The formats for representing core resources must be independent of the individual technologies that manage them, ensuring minimal lock-in and greater agility in extending the platform in the future.

Summary

In this chapter, we explored how to support the lifecycle of a data product in all its phases through an XOps platform designed to mobilize the entire data ecosystem in managing data as a product.

First, we delved into the role of the XOps platform within the data ecosystem. We saw how, through its value production engines (transaction engine and learning engine), it reduces friction in the creation of data products, computational policies, and ontologies, facilitates their operational management, and enables effective and efficient use of data to support the organization's priority business cases.

We then examined the logical architecture and critical capabilities offered by the XOps platform. Specifically, we analyzed how these capabilities are orchestrated by the XOps platform to support the development, operation, and consumer experience.

Finally, we explored how to implement the XOps platform incrementally as a composition of modules, proposing key architectural principles to guide the make-or-buy decision for each module.

In the next chapter, we will begin to explore strategies for defining a path to adopting a data-as-a-product management paradigm.

Further reading

For more information on the topics covered in this chapter, please see the following resources:

- *Platform Revolution: How Networked Markets Are Transforming the Economy—and How to Make Them Work for You*, G. Parker, M. W. Van Alstyne, S. P. Choudary (2016) `https://www.amazon.com/gp/product/0393249131`

- *Platform Strategy: Innovation Through Harmonization*, G. Hohpe, M. Danieli, JF Landreau (2024) `https://www.amazon.com/gp/product/B0D1R6DX2M`

- *Platform Design Toolkit* – Boundaryless (2013) `https://www.boundaryless.io/pdt-toolkit/`

- *Data Developer Platform* `https://datadeveloperplatform.org/`

- *Platform as a Product*, M. Pais (2022) `https://www.youtube.com/watch?v=b8YHCDMxqfg`

- *Open Data Mesh Platform* `https://platform.opendatamesh.org/`

Get This Book's PDF Version and Exclusive Extras

UNLOCK NOW

Scan the QR code (or go to `packtpub.com/unlock`). Search for this book by name, confirm the edition, and then follow the steps on the page.

Note: Keep your invoice handy. Purchases made directly from Packt don't require one.

Part 3:
Designing a Successful Data Product Strategy

In this part, we'll explore how to design and implement an incremental, value-driven data strategy for successfully adopting the managing data as a product paradigm. We'll analyze key elements of the strategy and how to define them in the early stages of the adoption journey. Understanding how to kickstart the initiative and secure buy-in from key stakeholders will be covered. Subsequently, we'll examine how to scale adoption across the enterprise by fostering a data-driven culture based on the principle of managing data as a product. Finally, we'll delve into data modeling in a distributed environment, emphasizing how modeling, both at the physical and conceptual levels, is a crucial element for fully leveraging the potential offered by modern generative AI solutions.

This part has the following chapters:

- *Chapter 8, Moving through the Adoption Journey*
- *Chapter 9, Team Topologies and Data Ownership at Scale*
- *Chapter 10, Distributed Data Modeling*
- *Chapter 11, Building an AI-Ready Information Architecture*
- *Chapter 12, Bringing It All Together*

8

Moving through the Adoption Journey

In the previous chapter, we saw how to implement an XOps platform to automate the data product life cycle. In this chapter, we will see how to successfully move through the adoption journey of the managing data as a product paradigm.

We will first analyze how a new paradigm spreads within an organization by introducing the main phases of the adoption journey.

Next, we'll delve deeper into each phase, exploring its objectives, the unique challenges it presents, and the critical organizational and operational activities required at each step.

Finally, we will see how to define a data strategy that can evolve agilely in the different phases of the adoption journey, adapting to the needs of each of them, and making full use of what has been learned in the previous phases.

This chapter will cover the following main topics:

- Understanding the adoption phases
- Delving into the assessment phase
- Delving into the bootstrap phase
- Delving into the expansion phase
- Delving into the sustain phase
- Driving the adoption with an adaptive data strategy

Understanding adoption phases

Shifting to managing the data-as-a-product paradigm is a transformative journey that ripples across the entire organization. It goes beyond operational changes, impacting both organizational structures and company culture. Such a change cannot happen overnight. It takes time and must be done in an incremental manner. In this section, we will explore the dynamics and main phases that characterize the process of adopting a new paradigm and how to prepare to embark on this challenging journey of transformation.

Tracing the journey ahead

Everett Rogers' **diffusion of innovation theory** sheds light on how new ideas, such as managing data as a product, take hold within organizations. Rogers proposes that innovations spread gradually through social systems, which can encompass communities, organizations, or any interconnected group. This gradual adoption stems from the varying openness to change among individuals within the system. Rogers identified five categories of adopters based on their inclination to embrace new ideas:

- **Innovators**: The thrill-seekers, eager to try new things and be the first
- **Early adopters**: The respected trendsetters, influencing others through their adoption
- **Early majority**: The cautious pragmatists, needing to see the innovation's value before adopting
- **Late majority**: The skeptical followers, adopting only after widespread acceptance
- **Laggards**: The tradition-bound resisters, needing strong evidence to overcome their aversion to change

The distribution of individuals within a social system according to their openness to change and innovation generally follows a bell curve, with two-thirds of the population being moderately in favor or moderately against change, while the remaining third is evenly divided between those highly in favor of change and those extremely opposed to it. The innovation to be introduced into the system must progressively and incrementally win over all categories of people, starting with those most open to change (innovators) and proceeding in an orderly manner from category to category until reaching the most resistant (laggards). The adoption of an innovation over time is therefore given by the cumulative distribution of the population in the different adopter categories (*Figure 8.1*).

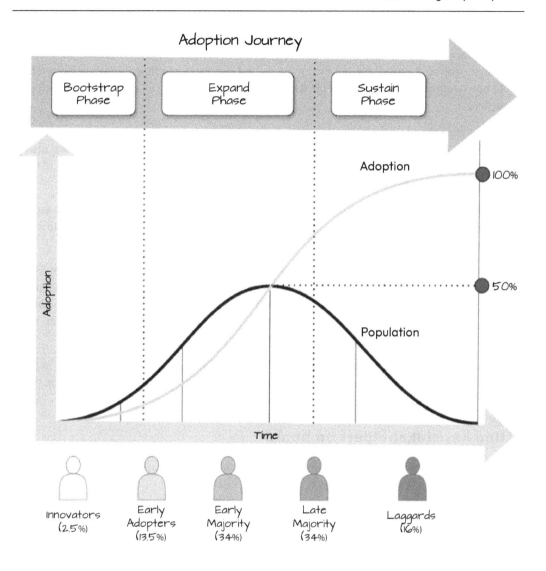

Figure 8.1 – Diffusion of innovations

This S-shaped curve, also known as the **logistic curve**, plots the number of adopters on the vertical axis and time on the horizontal axis. By analyzing this curve, we can divide the adoption journey of a new idea into three main phases: bootstrap, expand, and sustain.

During the **bootstrap phase**, adoption is slow and limited to innovators and early adopters. The new idea is still in its early stages and needs to be evaluated in a specific context. The level of uncertainty is high, and the application is still experimental, relying heavily on customized solutions and evolving practices.

If the bootstrap phase proves successful and validates the new idea's value in the given context, a transition to the expand phase can begin. In the **expand phase**, the adoption of the idea is rapidly scaled across the population, starting with the early majority and eventually reaching the late majority. The level of uncertainty decreases, and the application of the idea starts to rely on standardized solutions and practices.

By the end of the expand phase, most of the population has adopted the new idea and experienced its value. This allows for a transition to the **sustain phase**, where even laggards adopt the new idea, making it a part of business as usual. The level of uncertainty is minimal, and the application of the idea is based on commoditized solutions and practices within the organization.

Before embarking on the three phases of adopting a new idea, a preliminary **assessment phase** is essential. During this phase, the idea is thoroughly evaluated to determine whether and how to proceed with its adoption.

Each phase in the adoption journey, from the assessment phase to the sustain phase, must be completed in a sequential manner. Each phase presents unique objectives, challenges, and demands, necessitating distinct adoption strategies and techniques. The most challenging aspects of the adoption journey often lie in the transitions between phases, where strategies, techniques, and sometimes even the individuals or teams leading the change initiative must adapt and evolve.

Before we dive deeper into each phase, its goals, and the key challenges associated with them in the coming sections, let's address in the next section a crucial element for successfully embarking on this journey: securing the involvement of key stakeholders.

Getting key stakeholders on board

A paradigm shift is a profound change within the organization. It is unlikely to be successfully completed following a bottom-up approach. Strong sponsorship from the top is required. Securing the sponsorship from the head of data management (i.e., CDO, CIO, etc.) is crucial for embarking on a journey to start managing data as a product. If, in addition to managing data as a product, it is also necessary to decentralize data ownership to the business domains from the outset, then a higher level of sponsorship may also be required outside of the data management function.

A common mistake in this type of transformational initiative is to start with ambitious goals that are not supported by the right level of sponsorship. The result is a failure to move beyond the bootstrap phase and scale the adoption of the new paradigm across the rest of the organization. Since the data-as-product paradigm acquires value through network effects as more data products are developed across the organization, if the adoption cannot scale beyond the perimeter of innovators and early adopters, it is doomed to fail.

While a bottom-up approach alone rarely drives a full-scale paradigm shift, individuals operating closer to the edges of the organization can still try to create an initial spark by engaging innovators and early adopters. By co-creating with these pioneers, they can build and deliver small, focused data products that demonstrate the tangible value of this new approach. These early, localized successes can serve as a beacon, illuminating the potential for broader organizational impact. As these early adopters generate momentum, their results create a powerful narrative, making it easier to gain executive sponsorship and catalyze the necessary top-down push for scaling the data-as-product paradigm.

However, having the right level of sponsorship is not enough. It is also necessary to obtain the buy-in of key people within the various categories of adopters as adoption scales. In other words, an adoption process is needed in which the new paradigm spreads by direct transmission from one category of adopters to another (inside-out approach) and not by forced co-option (top-down approach).

The assessment phase serves as a critical initial step to verify the feasibility of adopting the new paradigm. This phase evaluates the adequacy of organizational sponsorship and buy-in from key stakeholders who will be involved in the next phases. Let's see how in the next section.

Delving into the assessment phase

The assessment phase is composed of a series of workshops aimed at generating alignment of vision among the key stakeholders impacted by the new paradigm on why it is valuable and what is needed to adopt it. In this section, we will explore how these workshops can be organized based on the specific deliverables that are intended to be produced as a result of the assessment phase.

Preparing for the assessment

The key outcomes of the assessment phase include the following:

- A common understanding among stakeholders of the new paradigm and its goals
- A shared and thoughtful awareness on why the new paradigm brings value and makes sense in the given context
- The establishment of an operating model to govern the adoption process
- A roadmap outlining the next steps and key activities, both technological and organizational, for the expand phase

The duration of the assessment phase varies depending on the scope of the initiative and the level of detail desired for the deliverables produced after the assessment. However, the assessment phase must absolutely be time-boxed. This means that it is important to set a fixed time frame, which can typically range from two days to two months depending on the case, and to organize accordingly to produce the deliverables within the established deadline. The goal is to decide whether and how to proceed, not to define an adoption plan at the highest level of detail to be executed in waterfall mode over the years to come. Any decision made in this phase can still be modified in the future based on what is learned along the way.

Not having fixed time limits for this phase generally leads to two possible problems: **loss of effectiveness** or **over-analysis**.

Loss of effectiveness occurs when, in the absence of precise deadlines, the people involved participate in assessment activities in a best-effort manner. This leads to a loss of focus and, consequently, effectiveness. It is better to reduce the number of meetings and the expectations on the level of detail of the deliverables, but to ensure that all stakeholders have the time to participate continuously and actively in the scheduled assessment meetings.

Over-analysis occurs when new discussions are continually opened in an attempt to find an answer to all the possible doubts, problems, or criticalities that may arise. It is better to accept that not everything can be foreseen in advance and that it is generally better to postpone some decisions until more information has been gathered rather than to invest unnecessary time into an analysis that, no matter how detailed, will never be complete.

Once the number of meetings has been determined, it is necessary to identify the right people to involve. In general, in addition to the sponsor and their direct reports, it is useful to invite the following people:

- **Senior leaders above the sponsor**: To make them aware of the initiative, obtain their feedback, and possibly secure their sponsorship too

- **Key business users**: To gather ideas and feedback from individuals across departments who rely on data to make informed decisions and will directly benefit from the adoption of the new paradigm

- **Key people involved in the bootstrap phase**: To make them actively involved in decisions that will affect them in later adoption phases

- **Subject matter experts** (SMEs): To identify early potential issues related to security, compliance, technology constraints, resource allocation, and so on

The main sponsor is responsible for scheduling the meetings, inviting each one to the right people, and ensuring their attendance. In general, the meetings are divided into two parts. The first part focuses on understanding and analysis, while the second part is dedicated to synthesis and deliverables production. The meetings and activities of each part are shown in *Figure 8.2*.

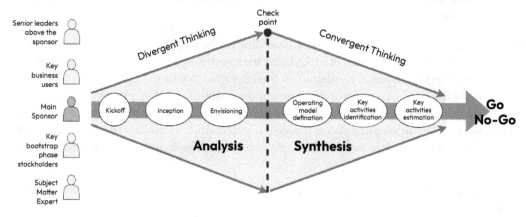

Figure 8.2 – Assessment phase structure

Analysis meetings delve into the reasons behind adopting the new paradigm, while synthesis meetings focus on how to manage the adoption process. The specifics of what needs to be done are addressed in subsequent phases. Let's now examine the structure of meetings and activities of the analysis part of the assessment phase.

Understanding the why

The analysis part of the assessment starts with a *kick-off meeting*. During the kick-off meeting, the main sponsor introduces the initiative, explains the objectives, and outlines the structure of the assessment. By the end of the kick-off meeting, each participant should have a clear understanding of the assessment goals, the upcoming activities, and their specific role in the assessment.

After the kick-off meeting, *inception meetings* follow. Inception meetings focus on analyzing the current issues within the data management function in relation to the support it provides to the business strategy. Before identifying these issues, it's important to define the initiative's scope, which will guide the analysis. Generally, a paradigm shift encompasses the entire organization. However, in large organizations, some parts may be excluded, or the scope may be limited to a specific branch.

Once the scope is defined, a brainstorming session can be held to identify the main issues affecting the data management function within this scope. These issues are then grouped by type (e.g., delivery speed issues, lack of necessary data, etc.). Each group is associated with one or more needs that the data management function is expected to meet in the future. For each need or group of needs, goals are defined, which can later be evaluated using specific metrics to assess how new initiatives promoted by the data management function will support the organization's needs.

By the end of the inception meetings, participants should have a strategic alignment on the key objectives that the data management function must achieve to support the needs expressed by various stakeholders. It is crucial that the needs determining the goals are directly linked to the business strategy and not abstract. For this reason, it's extremely important for key business figures to participate in these assessment meetings.

The information gathered during the inception meetings can generally be compiled into a **vision board**, as shown in the following figure (*Figure 8.3*).

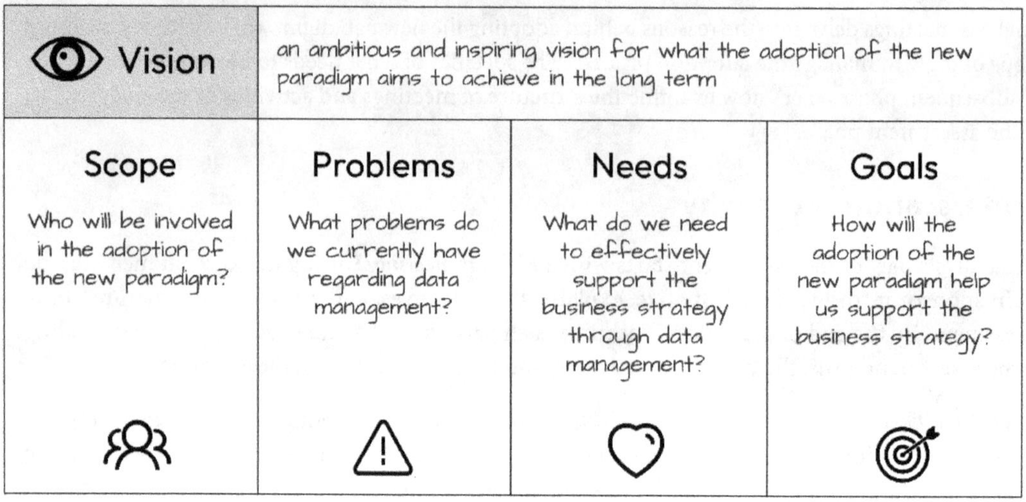

◉ Vision	an ambitious and inspiring vision for what the adoption of the new paradigm aims to achieve in the long term		
Scope	**Problems**	**Needs**	**Goals**
Who will be involved in the adoption of the new paradigm?	What problems do we currently have regarding data management?	What do we need to effectively support the business strategy through data management?	How will the adoption of the new paradigm help us support the business strategy?

Figure 8.3 – Vision board

After the inception meetings are concluded, the analysis moves on to the envisioning meetings. Envisioning meetings evaluate how the new paradigm centered on managing data as a product can support the data management function in achieving the goals identified during the inception meetings. In the first envisioning meeting, the main sponsor or their direct reports present the new paradigm, its principles, and the key capabilities that need to be implemented for it to be effective. With a shared understanding of the new paradigm, its principles and capabilities are analyzed to identify, on one hand, how they can support the data management function in reaching the identified goals compared to the current approach (opportunities), and on the other hand, the challenges of implementing them in the given context (threats).

During the envisioning phase, in addition to evaluating the adequacy of the new paradigm concerning the goals and the organization's readiness to adopt it, alternative approaches can also be comparatively assessed. This provides more elements for making an informed decision on whether or not to proceed with the adoption. By the end of the envisioning meetings, participants should clearly understand why it makes sense to apply the new paradigm or not, based on which assumptions, and with what potential challenges in adoption.

The information gathered during the analysis conducted in the envisioning meetings can generally be compiled into a **Risks, Assumptions, Issues, and Dependencies (RAID) matrix** as shown in the following figure (*Figure 8.4*).

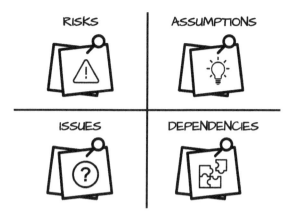

Figure 8.4 – RAID matrix

After completing the analysis meetings, an initial checkpoint can be conducted to evaluate whether it makes sense to proceed with the synthesis meetings. At this stage, it may become clear that the data-as-a-product paradigm cannot currently be adopted. Reasons for this could include a lack of sufficient sponsorship, a preference for an alternative approach among innovators and early adopters, or significant gaps and blockers that need to be addressed before reconsidering the adoption of the new paradigm.

Let's see now the structure of synthesis activities and meetings.

Defining the how

Following the analysis meetings, if the decision is to move forward, the first step involves translating the goals identified during the inception meetings into concrete, measurable indicators known as **Measures of Success (MoS)**. This allows for measurable evaluation of how much the adoption of the new paradigm impacts the organization's needs. MoSs are essential for creating strategic alignment, defining actions, and measuring outcomes. They are also necessary to establish the operating model adopted for the bootstrap phase and then evolve in subsequent phases. This operating model will guide the identification of high-impact activities to be undertaken in each iteration and how to evaluate their results against the defined MoS.

Multiple methodologies define explicit success metrics and then use and evolve them over time to ensure constant and continuous alignment between strategy and execution. These methodologies can be used as a basis for defining the operating model to use. In this chapter, we will use the EDGE methodology, which we will describe in the *Governing the adoption process with an adaptive data strategy* section. Other popular methodologies are **Objective and Key Results (OKR)**, **Gartner DASOM**, **balanced scorecard**, **impact mapping**, and so on. In the *Further reading* section, there are links to some resources for those who would like to learn more about these alternative methodologies.

If success metrics indicate the direction, the operating model is the steering wheel that guides the organization along the adoption journey. However, to start the engine, you need gasoline. To this end, it is necessary, in the synthesis part of the assessment, to define the main initiatives that will be carried out in the bootstrap phase and to secure the necessary budget to implement them. In particular, it is necessary to define the first business cases that will be implemented using the new paradigm (**lighthouse projects**) and the main initiatives that will be rolled out to support them.

The business cases generally already have a budget allocated. However, implementing them using the new paradigm in the bootstrap phase has extra costs that will only be amortized in subsequent phases. It is therefore necessary to secure the necessary budget to carry out the exploratory activities related to the adoption of the new paradigm. The idea of being able to drown these activities in the only budget allocated to the lighthouse projects is an illusion. Exploratory activities lay the foundation for the success in scaling adoption in subsequent phases. They cannot be done in best-effort mode.

At the end of the assessment, all the necessary elements are available to make a shared and informed decision on the proposal to adopt the new paradigm. If the choice is positive, it is possible to proceed with the bootstrap phase, which we will discuss in more detail in the next section.

Delving into the bootstrap phase

The bootstrap phase aims to lay the foundation for the adoption journey. In this section, we will see what the main activities that characterize this phase from both an organizational and operational point of view are.

Setting the foundation

The bootstrap phase begins by establishing organizational structures necessary to execute the operating model defined during the assessment phase. The operating model dictates decisions about activities throughout the bootstrap phase and beyond, guiding prioritization, budgeting, and the ongoing monitoring of identified tasks. Consequently, it is crucial to reorganize roles and management responsibilities within the data management function (System 2-5) to ensure alignment with the chosen operating model before moving forward.

As discussed in *Chapter 3, Data Product-Centered Architectures*, the data management function to adopt the data as a product paradigm needs to implement the following four key capabilities at the operational level (System 1):

- Data product development
- Governance policy-making
- XOps platform engineering
- Data transformation enabling

To achieve this, organizational changes within the operational plane of the data management function might be necessary. This could involve modifying the structure of existing teams and their responsibilities.

It is generally advisable to implement the needed organizational changes incrementally. For this reason, at the beginning of the bootstrap phase, it is often chosen to form a new team, or a limited number of new teams, tasked with initiating the adoption of the new paradigm according to the identified operating model. The pre-existing organizational structure remains unchanged and continues to operate as normal, managing all business-as-usual activities. Gradually, the newly formed team or teams start to specialize, structure themselves, and integrate into the existing organizational structure, modifying it from within. We will explore possible organizational structures in the next chapter (*Chapter 9, Team Topologies and Data Ownership at Scale*).

After making the necessary changes to the organizational structure of the data management function to execute the new operating model and implement the core capabilities required by the new paradigm, it is essential to define the initial concrete activities to begin the adoption journey. Specifically, it is necessary to identify the activities to implement each key initiative identified during the assessment phase. Based on the lighthouse projects to be developed following the new paradigm and the potential impact on success metrics, the identified activities should be prioritized and appropriately funded with a portion of the budget allocated to the initiative they belong to.

The planning of activities is governed by the control system (System 3) of the data management function. It is important that activity planning is managed in an agile way and conducted in short cycles. This way, it is possible to fully leverage what has been learned in previous iterations to guide future decisions and maximize the activities' impact on defined MoS. The higher the level of uncertainty is, the smaller the activities should be, and the more frequent the planning review meetings should be. This approach allows for experimentation while applying the new paradigm, taking controlled risks, and using organizational learning to quickly adjust as needed.

The activities identified within each initiative can be divided into streams based on the operational capability responsible for executing them (*Figure 8.5*).

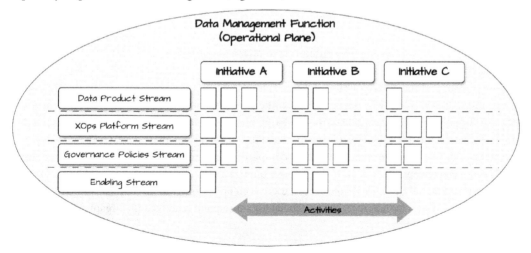

Figure 8.5 – Initiatives and activity streams

Initiatives and activities during the bootstrap phase vary depending on the specific context. However, it is generally possible to identify a minimal set of key activities that should be considered at this stage of the adoption journey for each capability. Let's see what these are, starting with the key activities to consider for the capability of data product development in the next section.

Building the first data products

The lighthouse projects selected during the assessment serve to test all the operational capabilities of the data management function envisioned by the new paradigm, not just those directly related to data product development. In fact, it is the lighthouse projects that influence the choice of which governance policies to define and which platform functions to implement first.

In general, it is always important to prioritize the identified activities related to the platform, governance, and enabling streams based on the data products that need to be developed to support the business cases of interest to the organization. This approach, in which the prioritization of activities is driven by business cases, ensures constant alignment between the needs of the business strategy and the needs of the data strategy based on the new data-as-a-product paradigm (*Figure 8.6*).

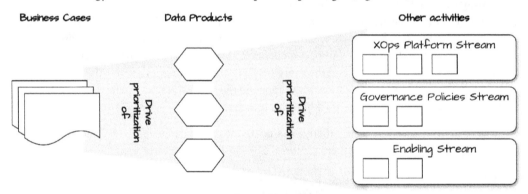

Figure 8.6 – Business case-driven prioritization

Given the impact that the choice of business cases has on all other activities related to adopting the new paradigm, it is extremely important to carefully select the business cases to use as lighthouse projects. Although the general context of the bootstrap phase is experimental, where progress is made through trial and error, it is still important to immediately tackle real business cases that bring value to the organization. The risks are obviously higher compared to experimenting with the new paradigm through proof of concepts. However, this approach offers a greater chance to learn and make quick adjustments. Additionally, in the case of success, there is an opportunity to create positive momentum toward adopting the new paradigm, which is crucial for subsequent scaling at the organizational level.

Choosing fake, overly simple, or low-value business cases can be a waste of time, while the opposite extreme can jeopardize the entire adoption process. Starting with overly complex business cases is risky because their intrinsic complexity can reduce the actual opportunities for experimentation. If successfully implemented, there may be few tangible advances in the adoption process. In case of failure, the new paradigm could easily be blamed, endangering future adoption phases.

Therefore, it is necessary to select business cases that are interesting for the business and sufficiently complex to serve as a valid test for the new paradigm, but not so complex as to risk failure from the outset. Business cases that require the development of source-aligned or consumer-aligned data products for a specific business domain are generally good candidates for lighthouse projects. On the other hand, business cases that depend on yet-to-be-developed infrastructural or platform components or require interaction between data managed by different domains can hide a high level of complexity and are generally not good candidates for lighthouse projects.

Although lighthouse projects are selected during the assessment phase, it is possible to consider changing them if, during the analysis aimed at identifying the necessary data products to support them, excessive complexities emerge. An example would be those mentioned earlier, which could pose too significant problems to solve in this initial phase.

Defining key governance policies

Governance policies serve to ensure the simple and secure composability of developed data products and to enable their automated management through standard services offered by the XOps platform. During the development of the first lighthouse projects, with a low number of data products already developed, it is not necessary to introduce too many governance policies. An excess of upfront policies could indeed prove to be unnecessary and slow down development. Policies should be defined incrementally, as needed, to address interaction, security, or automation issues.

Among the first governance policies to formulate are the definitions of what constitutes a data product, what metadata each data product must provide, and how this metadata should be represented through a descriptor document that can easily be accessed by potential consumers and the underlying platform. Specifically, clearly defining the metadata for annotating exposed data from the very beginning can help avoid challenges and delays later on, especially when users may struggle to categorize data retrospectively. It is also possible to define different types of data products and associate each type with different constraints on the metadata they must specify within the descriptor document.

As the bootstrap phase progresses, it may make sense to start defining standard APIs for some control, observability, and discoverability ports that all products must implement and expose. Standardizing these APIs is crucial for simplifying the operational management of products performed by the XOps platform.

Before completing the bootstrap phase and moving on to the expand phase, it is advisable to define governance policies aimed at standardizing the deployment process of data products to enable its orchestration and control by the XOps platform. Thanks to the standardization of the deployment process, the XOps platform can govern releases by enforcing shared rules through computational policies. This allows scaling the adoption by distributing responsibility across multiple autonomous teams while continuing to ensure the quality of released products.

All governance policies defined in this and subsequent phases should be documented through appropriate **Governance Decision Records** (**GDRs**) and made easily accessible to the entire organization. *Figure 8.7* shows an example of a GDR.

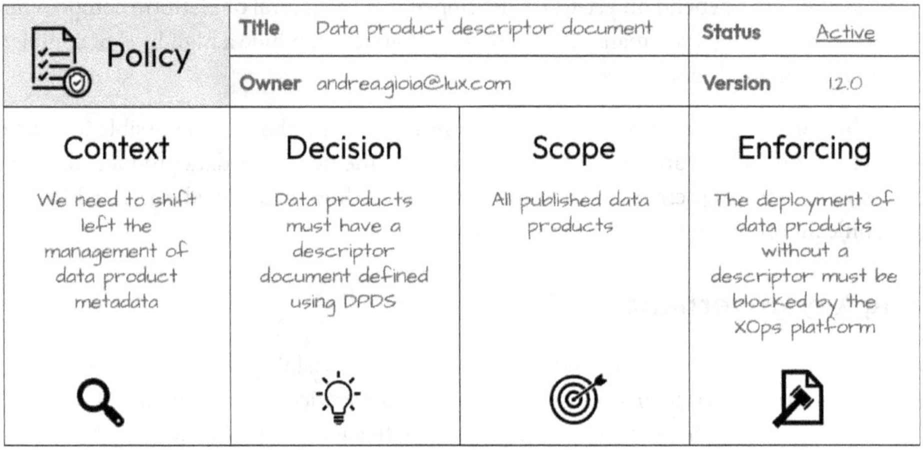

Policy	Title	Data product descriptor document	Status	Active
	Owner	andrea.gioia@lux.com	Version	12.0

Context	Decision	Scope	Enforcing
We need to shift left the management of data product metadata	Data products must have a descriptor document defined using DPDS	All published data products	The deployment of data products without a descriptor must be blocked by the XOps platform

Figure 8.7 – GDR example

Implementing the thinnest viable platform

The implementation of the XOps platform, whether through a make or buy approach, should be initiated concurrently with the first lighthouse projects.

The platform itself is a product that needs to evolve over time to support data products in compliance with governance policies. Although the benefits of the platform increase with the number of developed data products, it doesn't make sense to wait too long. Delaying the start of the platform's implementation wastes valuable time and exposes already-developed data products to potential rework. Similarly, waiting to develop the first data products until the platform is ready is extremely risky. This approach not only delays the first visible results for the business but also risks implementing platform functionalities that may not be needed. The best approach is to start immediately by implementing the **thinnest viable platform** with a few basic functionalities and then progressively add features as needs and problems arise during the development of data products.

The initial platform capabilities may vary depending on the specific context and the selected lighthouse projects, which will help guide the prioritization of key functionalities.

One of the first functionalities the platform can implement is a data product registry, as it is crucial for ensuring product discoverability. The registry is responsible for storing the descriptor documents of released data products and making them available to potential consumers. Regardless of how it's implemented, it's crucial that the registry allows potential consumers to easily find the available products through a self-service search process. In its simplest version, a data product registry can be implemented using a shared folder to save descriptor documents and a spreadsheet to catalog the names of data products and the references to the files containing their descriptors. More advanced versions can be implemented using custom services or by leveraging an external data catalog. On top of the registry, marketplace functionalities can later be implemented to facilitate the search and access requests for published data products by potential consumers.

Once the method for cataloging and making the descriptor documents of released data products easily accessible is defined, it is possible to start working on enforcing governance rules through computational policies. As discussed in previous chapters, a computational policy is a service that receives the resource to be validated through a standard API and returns the validation outcome. The XOps platform must provide the necessary functionalities to develop simple policy-as-code, for example, through adapters to external policy engines such as Open Policy Agent. It must also be able to manage custom policies when it is necessary to develop complex controls for which the expressiveness of external policy engines is insufficient. Finally, the platform must provide functionalities to catalog computational policies and manage their life cycle.

If, during the bootstrap phase, the governance capability standardizes the APIs of some product ports, it is useful to leverage these APIs to implement a control plane for product operations within the XOps platform. Through this control plane, it should be possible, for example, to monitor the status of a specific data product (integration with standard observability ports) and view or modify its internal configurations (integration with standard control ports).

Finally, in parallel with defining the policies that govern the deployment of data products, the platform could implement the necessary functionalities to orchestrate and monitor these deployments to act as a quality gate and ensure the compliance of released data products.

Concurrently with the implementation of the thinnest viable platform, it is also important during the bootstrap phase to address any gaps related to services or applications in the underlying infrastructure necessary for the development of new data products (e.g., configuring the data management landing zone, acquiring a new analytical database or a streaming platform, etc.).

Enabling the enablers

Even in the initial bootstrap phase, it is important to organize enabling activities to ensure that everyone involved has the same level of understanding of the new paradigm and how it is being implemented within the organization. At this stage, the focus is primarily on internal enabling for the teams involved in the ongoing adoption process, centered around study, discussion, and synthesis of relevant elements. In this regard, it can be useful to organize periodic meetings focused on discussing and analyzing specific topics of interest.

At the beginning of the bootstrap phase, the enabling capability is responsible for identifying topics that merit deeper exploration, finding interesting study resources, and organizing discussion and sharing meetings. Whenever possible, it is also extremely beneficial to have moments of interaction with external entities that have already embarked on this adoption journey. This can be done by participating in industry events or inviting external people to share their experiences in one of the internal discussion meetings.

As the bootstrap phase progresses, it is important to start transitioning from enabling within the teams directly involved in this phase to enabling towards external groups. In this sense, it is essential to identify the people who will be responsible for the enabling activities during the expand phase. These people are not necessarily the same ones who were most actively involved in internal enabling during the bootstrap phase. While internal enabling is entrusted to those most knowledgeable about the new paradigm and convinced of its benefits, external enabling requires patience and systematicity. It is not about discovering new content and proposing new ideas but about industrializing the onboarding and change management process necessary for scaling adoption. Not everyone who is well-suited to managing internal enabling may have the qualities to optimally manage enabling external groups.

Once the people responsible for external enabling are identified, it is necessary to start preparing for the expand phase. This includes beginning to produce a handbook that describes the main principles, processes, and procedures that guide activities within the context of the new paradigm. The handbook is a living document that evolves alongside the adoption process. The entire organization should have access to the handbook. It should be the primary reference for understanding the new paradigm and how it has been implemented for all individuals involved in its adoption.

Alongside the handbook, it may also be useful to prepare quickstarts or how-tos on handling common situations related to the development of new data products, their operational management, and the use of functionalities provided by the XOps platform. The objective of these documents is to formalize best practices and reduce the future workload on those responsible for enabling.

Transitioning to the next phase

The bootstrap phase can extend beyond the implementation of the first lighthouse projects if needed to consolidate processes, procedures, and platform functions necessary to support the expand phase. Tackling organizational scaling without having a solid foundation is very risky. However, it is important to set limits. Extending the bootstrap phase indefinitely makes no sense. Clear objectives for transitioning to the next phase and deadlines for achieving them must be defined. Generally, the bootstrap phase can last from 6 to 18 months. Exceeding two years indicates a lack of focus on the new initiative, insufficient sponsorship to scale adoption, or an overly cautious approach in the hope of resolving all potential problems in advance that might be encountered during the expand phase.

In the next section, we will continue to analyze the adoption journey of the new paradigm, focusing on the expand phase.

Delving into the expand phase

The goal of the expand phase is to scale the adoption of the new paradigm within the organization. In this section, we will explore the main challenges and activities that characterize this phase from both an organizational and operational perspective.

Crossing the chasm

The transition from innovators and early adopters (bootstrap phase) to the early majority and late majority (expand phase) in adopting the new paradigm is highly complex and marks a significant discontinuity in the adoption curve. This point, usually named "the chasm," represents the most critical juncture in the entire adoption journey, and the way it is managed generally determines whether the new paradigm will be adopted by the whole organization or not.

The difficulties in crossing this juncture arise from the significant differences between the adopter groups on the left side of the chasm and those on the right. Innovators and early adopters involved in the bootstrap phase are generally motivated and enthusiastic about the new paradigm. They are willing to accept risks and adopt evolving and imperfect solutions. The early majority, and even more so the late majority, are more pragmatic, seeking proven solutions that have demonstrated their value. They are less willing to take risks on something new and are naturally less inclined to change. Often, they are also tied to existing processes and solutions that they know well, and thus see the new paradigm as an unnecessary hassle imposed by the organization.

The bootstrap phase aims to reduce the risks associated with the new paradigm and present success stories of its application to encourage adoption by the early and late majority. However, this is only the starting point of the expand phase. In this phase, it is crucial to clearly incorporate the adoption of the new paradigm into the company's strategy, providing all involved organizational units with appropriate goals and incentives. Additionally, it is essential to continue investing in enabling activities to support change management and in the evolution of the XOps platform to minimize frictions and extra costs associated with the large-scale adoption of the new paradigm. We will explore how to achieve this in the following sections.

Scaling the adoption

Scaling the adoption within an organization can be done in different ways depending on the scope within which the new paradigm should be adopted. Specifically, the adoption process for managing scaling can vary significantly based on the level of decentralization in the ownership of data products. In a fully centralized data product management model, scaling the initiative primarily involves the IT teams responsible for the data function, with minimal involvement from external organizational units linked to various business domains. In a completely decentralized model, it is the opposite. Scaling the initiative mainly involves external organizational units, with only marginal involvement from IT functions. In any case, it is crucial to divide potential adopters into groups and approach scaling incrementally, focusing on homogeneous groups already present within the current organizational structure.

In a centralized model, the existing groups to onboard incrementally are IT teams. These teams are typically associated with specific technologies related to data value chain management. Here, the onboarding process aims to reorganize the activities and responsibilities of these teams to align them with business domains rather than technological components. This may require redistributing people among teams so that each team has the necessary skills to operate independently across all infrastructural technologies. The role of the XOps platform is crucial in reducing the complexities of the underlying technology, allowing these teams to focus more on understanding the business domains they are associated with under the new model than on technological problems. It is also important to ensure that the demand management process directs identified business cases to the appropriate teams based on the business domain to which the business case belongs.

In a completely decentralized model, the existing groups to onboard are those belonging to various organizational units associated with the company's business domains. The onboarding process aims to make them autonomous in producing new data products related to their specific domain. This may necessitate transferring people from central IT to the domains to distribute the technical skills required for product development. The XOps platform remains fundamental in reducing the complexity of the underlying technology, enabling domain teams developing data products to rely less on highly specialized technical people for their activities.

In the case of hybrid models, where some business domains take responsibility for developing their products while others continue to rely on central IT, the two adoption strategies described must be combined and executed in parallel.

Evolving governance policies and platform services

It is important to extend governance and control functions to include representatives from new teams adopting the new paradigm during scaling, ensuring their involvement in decisions that will directly impact them operationally.

Additionally, it is crucial to define new policies to ensure the control of data products not only during deployment but also post-release.

Since more data products from different domains are being released during this phase, the governance capability must also be organized to ensure their interoperability, not only technologically but also semantically. This involves working on methods to simplify semantic reconciliation between data exposed by different domains. We will explore this topic further in *Chapter 10, Distributed Data Modeling*.

Besides governance, the platform's capability is extremely important during this phase. The XOps platform, in particular, should focus on simplifying the development of new products through the definition of building blocks, blueprints, and sidecars that can be easily composed and configured.

Increasingly, the monitoring and observability functionalities for released data products also become important to promptly highlight issues and promote trust in the quality and availability of released data products.

Dealing with legacy systems

In the expand phase, unlike in the bootstrap phase, it is common to have to deal with legacy data management systems (e.g., old data warehouses or data lakes). Unless there are urgent constraints necessitating the migration of these legacy systems to the new data-as-a-product paradigm (e.g., end of support for underlying technology, security or scalability issues, etc.), it is always better to avoid a big-bang migration. An incremental approach driven by actual business needs is preferable.

First, it is advisable to isolate these systems using a series of **pseudo-data products**. A pseudo-data product exposes some of the data contained in the legacy system in a manner that aligns with the new paradigm (i.e. using ports, descriptor documents, etc). All downstream data products that need to access data managed by the legacy system do so through these pseudo data products, rather than directly accessing the legacy system. A pseudo-data product is essentially similar to a source-aligned data product but does not require moving data from the legacy system. It simply provides an abstraction layer that mediates access to the legacy system. It is important that these pseudo-data products are clearly identified as such. In fact, while they are indistinguishable from other data products from a consumption perspective, they do not have the same capacity to evolve agilely over time.

Once the legacy system is isolated, it can undergo incremental refactoring using the **strangler fig pattern**. In essence, each time a new business case requires modification of a pseudo data product, it provides an opportunity to transform it into a real data product by transferring the data management logic from the legacy system to the new data product. This way, no new development occurs on the legacy system. New business cases incrementally drive the refactoring of the legacy system, which will be completely emptied over time and can then be decommissioned.

In the next section, we will continue to analyze the adoption journey of the new paradigm, focusing on the sustain phase.

Delving into the sustain phase

The sustain phase aims to finalize the adoption of the new paradigm by industrializing practices and processes to maximize the benefits of economies of scale. In this section, we will explore the main challenges and activities that characterize this phase from both an organizational and operational perspective.

Becoming the new normal

In the sustain phase, the new paradigm has been adopted by most of the organization. The adoption process is now well-established and irreversible. At this stage, the focus shifts from increasing adoption to leveraging the economies of scale offered by the adoption achieved in previous phases.

There are two main areas of efficiency improvement in this phase that can maximize the benefits of having adopted the new paradigm: intelligent automation and user experience enhancement.

Intelligent automation focuses on using AI techniques to further automate the development and operational management of data products. Thanks to intelligent automation, it is possible to reduce development time and management costs for data products. We will delve deeper into the use of AI in data product management in *Chapter 11, Building an AI-Ready Information Architecture*.

User experience enhancement, on the other hand, focuses on leveraging the existing portfolio of data products to simplify their use and composition for developing new business cases. In particular, during this phase, the introduction of low-code/no-code interfaces can be evaluated to allow business users to independently compose existing data products according to their needs. This simplifies data access and utilization within the organization, eliminating the need for external support to implement data solutions. As a result, response times are accelerated, development costs are reduced, and business agility is enhanced, enabling various business functions to quickly leverage data and better adapt to the changing needs of their operating environment.

Remaining capable to adapt

Industrializing services and processes to leverage economies of scale and achieve operational efficiency should not interrupt the process of continuous exploration and improvement. Without a constant push in this direction, there is a risk of losing agility and adaptability in the medium term. Therefore, the exploitation of the new paradigm must always be appropriately balanced with investments in organizational and technological innovation.

In essence, industrializing services and processes and promoting innovation are not opposing objectives, but rather two sides of the same coin. Finding the right balance between these two elements is crucial for achieving a sustainable competitive advantage in data management.

Thus, we have completed the journey through the various phases of the adoption process. In the next section, we will see how to define an adaptive strategy to successfully navigate through the different stages of this journey.

Driving the adoption with an adaptive data strategy

In this section, we will introduce the EDGE operating model, which we recommend using to define an agile, distributed, and adaptive data strategy to move successfully through all the phases of the adoption journey.

Understanding evolutionary strategy pillars

The data-as-a-product paradigm defines a set of connected principles, architectures, processes, and practices that must be contextualized during implementation based on the history and objectives of the specific organization. Therefore, it is not possible to apply this paradigm uniformly in every context.

Thus far, we have described the adoption journey as a sequence of phases, specifying objectives and challenges for each, but we have only provided high-level suggestions regarding the concrete activities to be performed in each phase. It is the responsibility of each organization to define its own adoption journey incrementally, do experiments, learn from mistakes, and pivot when necessary.

It is evident that the adoption process is therefore based on continuous exploration and refinement, not on a precise plan that can be fully defined upfront during the assessment phase and then executed in a waterfall fashion until completion. Therefore, it is important to adopt an adaptive data strategy that can evolve over time based on needs and insights gained during the journey. To this end, it is necessary to reconsider the pillars of traditional strategic planning.

An adaptive strategy is not a well-defined plan that leads to achieving desired objectives, but rather a planning and results control model that ensures alignment at all times between strategic objectives and operational actions taken by the organization.

In summary, while a traditional strategy consists of a series of objectives and a plan to achieve them, an adaptive strategy consists of a series of objectives and an agile operating model designed to incrementally and possibly distributedly decide how to achieve them. Various operating models of this type exist in literature. In this section, we have chosen to present the EDGE model because we believe it is one of the best for bridging the gap between strategy and execution, responding quickly to changes, reducing risks, and maximizing organizational learning.

The EDGE model is based on three key elements: the **Lean Value Tree** (**LVT**), a value-based planning process, and a lightweight governance framework. We will delve into each of these elements in the upcoming sections.

Keeping strategy and execution aligned with EDGE

EDGE utilizes the LVT to ensure alignment between strategy and execution. At its root, an LVT outlines the strategic vision, which is then broken down into a series of goals. Each goal is linked to one or more initiatives that support its achievement in the LVT. Initiatives encompass portfolios of activities that explicitly outline what actions are taken to implement the initiative itself and, in doing so, move toward the direction articulated through the vision and its associated goals. Within the LVT, one can pinpoint what is being done at the execution level to achieve each strategic goal. Similarly, for every operational activity, it's possible to determine which strategic goal it supports (*Figure 8.8*)

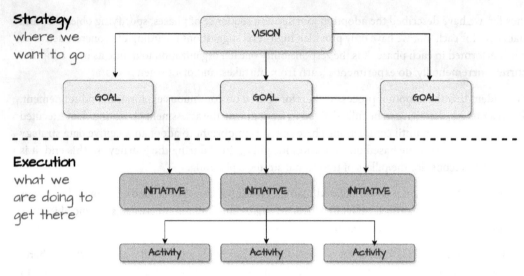

Strategy
where we
want to go

VISION

GOAL   GOAL   GOAL

- -

Execution
what we
are doing to
get there

INITIATIVE   INITIATIVE   INITIATIVE

Activity   Activity   Activity

Figure 8.8 – LVT

Therefore, the LVT serves as a tool for planning, communication, and operational monitoring, providing a clear view of an organization's strategic objectives and its progress toward achieving them. When properly defined, this tool bridges the traditional gap between strategic plans well understood by management and the decision-making processes of those involved in various operational activities.

The vision, goals, and initiatives of the data strategy are defined, if not already established, during the assessment phase. During this phase, it is ensured that the adoption of the new paradigm can effectively support the data strategy goals and, consequently, the overall business strategy. If the decision is made to proceed with the adoption journey, the activities associated with each initiative are then defined incrementally in the subsequent phases using the planning methodology defined by the selected operating model. Let's see how this is done within the EDGE operating model.

Planning the execution with EDGE

Each goal identified in the LVT must be linked with one or more MoS to objectively and collectively assess its achievement. The MoS may change over time, albeit infrequently. Operational activities, however, undergo frequent planning cycles. During each planning cycle, the impact of activities conducted in previous iterations on the associated MoS is assessed for each initiative. Decisions are then made on whether to halt, introduce new activities, or adjust priorities and budgets for ongoing ones. Prioritization is driven by the expected value each activity can deliver to the initiative's MoS (value score) and the effort required for implementation, calculated in terms of investment and risk (impact score). Activities with a higher potential value-to-effort ratio are given higher priority.

Each initiative has a dedicated budget, which funds its activities based on the priorities set during the planning phase (*Figure 8.9*)

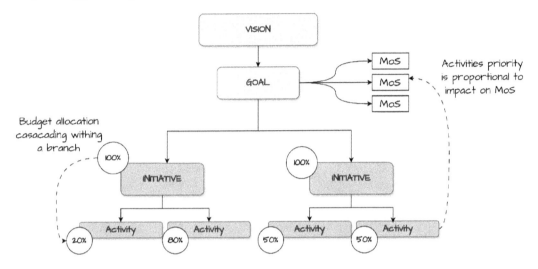

Figure 8.9 – Activities prioritization and budget allocation

These priorities and budgets typically can be modified at each planning cycle.

Evolving the strategy with EDGE

The LVT represents the current strategic plan but is designed to evolve continuously. Activities within initiatives can be changed during each planning cycle. Initiatives, goals, and MoS may also vary over time to adapt the data strategy to the changing business context. Generally, activities change on a monthly basis, while initiatives typically vary quarterly. Goals and MoS, on the other hand, usually change at most annually when it is needed to adjust the strategic direction. Finally, the vision should be stable over time, with changes occurring at most over multi-year cycles.

Moving through the LVT from root to leaves, the strategic vision is progressively translated into increasingly concrete and operational activities. The responsibility for defining and evolving the nodes of the LVT can be distributed across the organizational architecture, as discussed in *Chapter 3, Data Product-Centered Architectures*. Typically, senior management, specifically within the data management function in this case (System 2-5), defines the vision and goals. Meanwhile, operational teams (System 1) translate these goals into initiatives and activities (*Figure 8.10*).

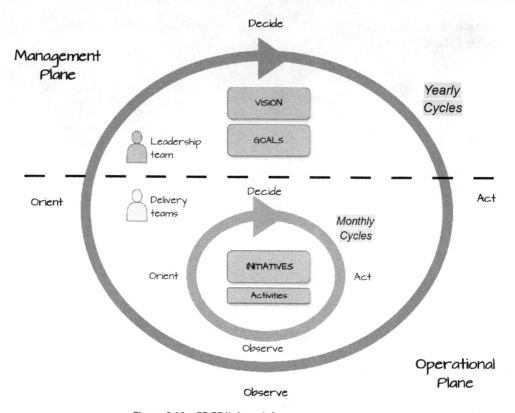

Figure 8.10 – EDGE lightweight governance process

The LVT and the lightweight governance process defined by the EDGE methodology to evolve over time allow for the implementation of an agile and adaptive data strategy. EDGE also facilitates the decentralization of strategy planning by empowering operational teams to actively participate in defining and refining strategies that align with overarching organizational goals. This decentralized approach ensures that strategic planning is not confined to a top-down directive but integrates insights and expertise from frontline teams who are closest to operational realities.

Monitoring the adoption process with fitness functions

The data strategy must be designed to support the overall business strategy. Objectives and associated MoS should be linked to the organization's strategic vision and business goals, and must therefore be easily understood by the business. It is generally incorrect to define the adoption of a technology or paradigm as a goal of the data strategy. Technologies and paradigms are tools that help achieve strategic goals; they should not be strategic goals themselves. The selection and adoption of technologies and paradigms consists of activities or groups of activities within the LVT. The LVT enables the selection of the most effective technologies and paradigms to support strategic goals and monitors their impacts during their adoption process.

The LVT can demonstrate to the organization that adopting the data-as-a-product paradigm is helping to achieve its strategy. Additionally, the LVT guides activities related to adopting the new paradigm, ensuring that these activities are not ends in themselves but always have positive impacts on the organization. The following figure shows the LVT representing the data strategy defined by LuX, the retail company we introduced as an example in previous chapters (*Figure 8.11*).

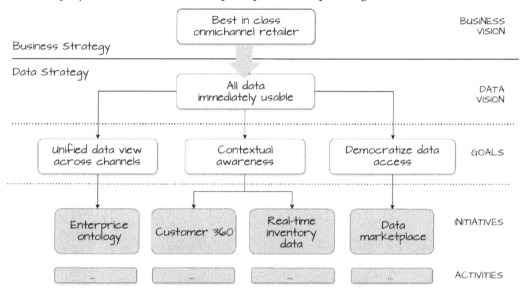

Figure 8.11 – The LuX LVT

However, the LVT does not immediately provide evidence of the progress being made in the adoption of the new paradigm. To this end, it is necessary to complement the MoS with specific measures to evaluate how well the adoption status of the paradigm aligns with the desired end-state of the socio-technical architecture needed to support the data-as-a-product paradigm. These measures are known as **fitness functions**.

Generally, fitness functions can be grouped according to specific areas of the architecture they assess. In the case of an architecture centered on data-as-a-product management, it is beneficial to group the defined fitness functions by the core capabilities of the data management function required by the new paradigm.

While goals and MoSs vary based on strategic business needs, fitness functions change according to the requirements of the specific phase of the adoption journey. As shown in *Figure 8.12*, MoSs monitor the alignment between activities and business objectives (e.g., how many users have access to up-to-date customer data?), while fitness functions monitor the adoption process of the new paradigm (e.g., how many released data products are actively used in more than one business case?).

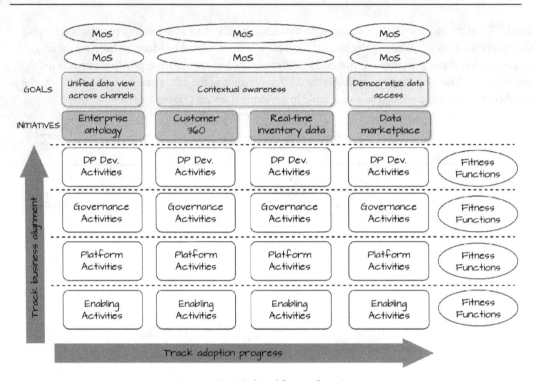

Figure 8.12 – MoS and fitness functions

Let's summarize the chapter now.

Summary

In this chapter, we saw how to successfully move through the different phases of the adoption journey of the data-as-a-product paradigm.

Firstly, we introduced Rogers' theory on the diffusion of new ideas within an organization. Using this theory, we defined a journey for adopting the data-as-a-product paradigm consisting of four main phases: assessment, bootstrap, expand, and sustain.

Next, we have detailed each of these four phases, discussing their objectives, challenges, and the key operational and organizational activities required.

Lastly, we introduced the EDGE operating model and illustrated how to use it to craft an adaptive data strategy capable of evolving over time to meet the specific needs of each phase of the adoption journey.

In the next chapter, we will explore how to structure the key teams involved in adopting the new paradigm and manage their responsibilities and interactions effectively.

Further reading

For more information on the topics covered in this chapter, please see the following resources:

- *Diffusion of Innovations* – E. Rogers (2003) `https://www.amazon.com/Diffusion-Innovations-5th-Everett-Rogers/dp/0743222091`

- *The Product Development Triathlon* – K. Beck (2018) `https://medium.com/@kentbeck_7670/the-product-development-triathlon-6464e2763c46`

- *Data Mesh Accelerate Workshop* – P. Caroli, S.Upton (2023) `https://martinfowler.com/articles/data-mesh-accelerate-workshop.html`

- *Pioneers, Settlers and Town Planners* – S. Wardley (2012) `https://blog.gardeviance.org/2012/06/pioneers-settlers-and-town-planners.html`

- *Crossing the Chasm* – J. Moore (2014) – `https://a.co/d/0b0rw20l`

- *Radical Optionality* – M. Reeves, M. Moldoveanu, A. Job (2023) `https://hbr.org/2023/05/radical-optionality`

- *EDGE: Value-Driven Digital Transformation* – J. Highsmith, L. Luu, D. Robinson (2019) `https://a.co/d/09kW9pg2`

- *Measure What Matters: How Google, Bono, and the Gates Foundation Rock the World with OKRs* – J. Doerr, L. Page (2018) `https://a.co/d/0hnM68Hh`

- *The Balanced Scorecard: Translating Strategy into Action* – R. S. Kaplan, . David P. Norton (1996) `https://a.co/d/09S8dZol`

- *Impact Mapping: Making a big impact with software products and projects* – Gojko Adzic (2012) `https://a.co/d/0d62JV6V`

Team Topologies and Data Ownership at Scale

In the previous chapter, we unpacked the different stages that are involved in the shift toward managing data as a product. We also explored how to craft an adaptive data strategy to navigate these stages seamlessly, ultimately achieving a company-wide adoption of this new data management paradigm.

In this chapter, we'll learn how to design an optimal organizational structure so that we can implement the socio-technical architecture of the data management function described in *Chapter 3, Data Product-Centered Architectures*.

Specifically, we'll explore how to organize operational and management teams and their interaction modes based on the organization's context and goals to achieve a fast flow of data product-driven value delivery.

This chapter will cover the following main topics:

- Introducing Team Topologies
- Defining operational teams
- Defining management teams
- Evaluating decentralization strategies

Introducing Team Topologies

The organizational architecture is built upon teams as its fundamental building blocks. These teams translate the key functions and capabilities identified in the architecture into reality. In this section, we'll delve into the core concepts of Team Topologies, a team-centered framework for organizational design defined by Matthew Skelton and Manuel Pais. Then, we'll leverage this framework throughout this chapter to describe how to optimally structure teams within the data management function so that we can manage data as a product.

Dissecting organizational architecture

An organization is a complex socio-technical system. Organizational architecture describes how the social part of this system is structured. Specifically, organizational architecture describes how an organization instantiates and governs the capabilities necessary to implement the key value streams for the realization of its business model. Organizational architecture consists of three building blocks: the organizational chart, team structure, and operating model.

The **organizational chart** describes how responsibilities for individual capabilities are redistributed within the organization. Typically, the organizational chart takes the form of a hierarchical chart, similar to the one shown in *Figure 9.1*:

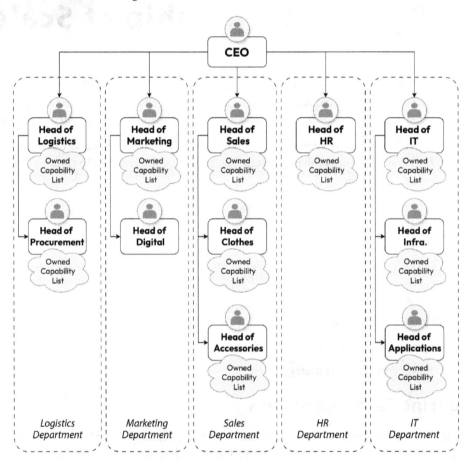

Figure 9.1 – Organizational chart

The organizational chart is designed and evolved to maximize the effectiveness of implementing and managing key capabilities. The focus is on optimizing the allocation of objectives, resources, and control structures to execute the organization's strategy. Therefore, the organizational chart emphasizes the formal, top-down reporting lines that govern these management aspects. However, it only offers a partial view of the organizational architecture since capabilities are interconnected and must interact horizontally, across organizational entities defined by the organizational chart, to implement the necessary value streams.

In this context, the **team structure** complements the organizational chart by defining which teams implement the individual capabilities and how these teams interact horizontally, either peer-to-peer or many-to-many, to ensure the organization's day-to-day operations. While the organizational chart focuses on the effectiveness of decision-making and resource allocation, the team structure focuses on the operational efficiency of implementing the value streams throughout the organization.

Teams are directly associated with nodes on the organizational chart, creating a comprehensive **organizational map** that represents both the formal top-down management structures needed for effective control and the informal, cross-functional value creation structures needed for efficient operations (*Figure 9.2*):

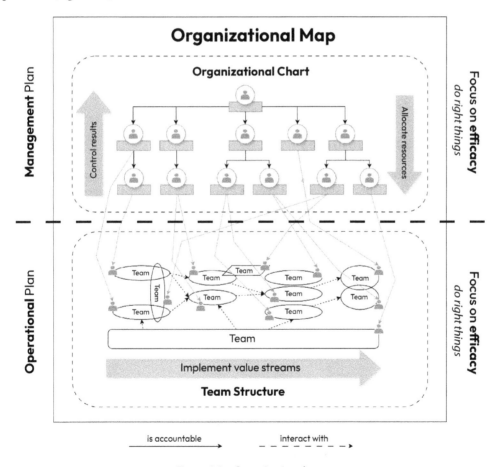

Figure 9.2 – Organizational map

The **operating model** complements the organizational architecture by procedurally describing how interactions between various organizational entities are governed. As we saw in the previous chapter, the operating model establishes how objectives are defined, how decisions on activity prioritization are made, how resources are allocated, and how the results that are achieved are controlled (*Figure 9.3*):

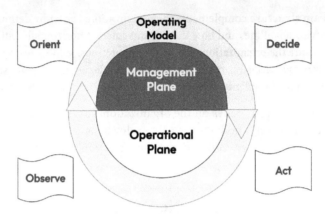

Figure 9.3 – Operating model

Now that we have an overview of the organizational architecture, we can delve deeper into the methods for defining teams and their interaction models.

Leveraging Team Topologies

Designing team structure is a key pillar of organizational architecture. The **Team Topologies** framework, developed by Matthew Skelton and Manuel Pais, provides essential patterns for designing team structures. The Team Topologies framework is based on the principle that function influences form and vice versa. This means that team form must be closely aligned with its functions (**Inverse Conway Maneuver**), while functions must also be defined carefully to minimize the cognitive load on each team. Form and function should be co-designed and evolved synergistically.

This key principle aligns perfectly with the paradigm of managing data as a product and the related socio-technical architecture that the data management function must adopt to implement it. For this reason, we'll introduce the Team Topologies framework in the following sections and use it throughout the rest of this chapter to understand how to structure teams within a data management function oriented toward managing data as a product.

Understanding team types

A team is a cohesive group of people who work together daily to achieve a common goal. Cohesion is what defines a team and enables it to function as a unit. This cohesion is built on mutual trust among its members. To achieve high levels of trust and, consequently, cohesion, a team needs to be small and stable.

The team should be small, generally no more than nine people, because it's difficult to establish high levels of trust with a large number of people. It should also be reasonably stable in its composition because building solid trust relationships takes time, typically from 3 weeks to 3 months or even longer. Therefore, it doesn't make sense to disband a team at the end of a project just when it's started to perform well.

Teams are the modular units of the organizational architecture. Each team aims to implement one or more capabilities necessary for the organization. The Team Topology framework identifies four types of teams: stream-aligned teams, enabling teams, platform teams, and complicated subsystem teams. Let's take a closer look:

- **Stream-aligned teams** are teams that are aligned with a single stream of work that delivers value to the organization. They have end-to-end responsibility for delivering a product or service and must be enabled to operate quickly and autonomously, minimizing external dependencies on other teams to achieve their goals. Stream-aligned teams are the primary type of team within an organization. They deliver direct value to the organization and are generally the most numerous. The other types of teams primarily aim to support stream-aligned teams by removing obstacles and simplifying their activities. These other teams provide value indirectly to the organization and are generally fewer in number.

- **Enabling teams** are designed to boost the autonomy of stream-aligned teams by helping them enhance their operational capabilities. These teams consist of specialists who work across various teams, dedicating their time to finding technical or process solutions for common issues that affect stream-aligned teams' operations. By doing so, enabling teams allow stream-aligned teams to continually improve their skills without losing focus on their main objective of delivering value to the organization. Enabling teams operate like internal consultants, providing guidance and transferring knowledge rather than doing the work themselves. Their role is to suggest ways to improve how activities are performed. An enabling team should quickly make a stream-aligned team self-sufficient in using a new solution that enhances its operational capacity so that it can assist another stream-aligned team or explore solutions for other operational challenges.

- **Platform teams** aim to increase the autonomy of stream-aligned teams by abstracting technological complexity through services or tools that assist them in their activities. The collection of tools and services provided by platform teams constitutes the platform(s) that stream-aligned teams can use to work more effectively, more quickly, and with fewer external dependencies. The success of a platform team is measured by the adoption of the platform they develop by the stream-aligned teams.

- **Complicated-subsystem teams** are responsible for handling complex capabilities that can't be easily managed by stream-aligned teams or simplified through services offered by platform teams. These capabilities may involve specialized areas of expertise (for example, security management, predictive model development, and so on) or require advanced knowledge of specific systems or technologies (SAP, other legacy systems, and so on). The primary goal of complicated-subsystem teams is to manage the complexity of these capabilities, providing support to stream-aligned teams as needed to help them perform related tasks effectively.

In addition to the team types defined by the Team Topologies framework, it's often useful to establish federated teams at the organizational level. A **federated team** isn't a traditional team in the usual sense. Its members are typically part of other teams. Each individual in a federated team usually spends only a small portion of their time on federated team activities, while the majority of their time is dedicated to their primary team. Federated teams generally fall into two categories: management committees and communities of practice:

- A **management committee** is a federated team whose members come together to make shared decisions that impact their respective primary teams. The adopted operating model defines the structure, responsibilities, membership, and decision-making processes of a management committee.

- A **community of practice** is a federated team that brings together individuals from different teams to facilitate the sharing of knowledge, experiences, and expertise on a specific topic of interest. Typically, people who perform similar activities in different teams may form a community of practice to learn from each other and exchange best practices. Communities of practice aren't an alternative to enabling teams. Enabling teams are cohesive groups that help teams improve their capabilities in an organized manner with clear objectives, one at a time. In contrast, communities of practice are open groups aimed at fostering organizational learning in a more spontaneous, bottom-up manner.

In general, federated teams can be seen as a specific subtype of enabling teams where members don't work full-time for the team. Due to their nature, it isn't suitable to federate other types of teams that are more focused on delivery activities. Each individual within the organization should always be a part of one primary team but can also belong to multiple federated teams.

To graphically represent the team structure, we'll use different shapes to identify each team, as shown in *Figure 9.4*:

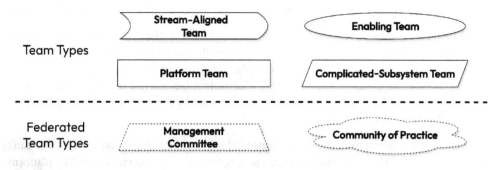

Figure 9.4 – Team types

Now that we've covered the main types of teams, let's look at how they can interact with one another.

Understanding interaction modes

Teams within an organization need to interact effectively to implement the value streams that define its business model. An organization is essentially a collection of interconnected teams working toward shared goals – a team of teams. When setting up the organizational structure with teams as modular units, it's important to focus not only on each team's scope and responsibilities but also on how they will collaborate. Poorly designed interaction modes are a major cause of friction and inefficiencies.

The Team Topologies framework identifies three possible modes of interaction between teams: collaboration, **X-as-a-Service** (**XaaS**), and facilitating.

Collaboration

Collaboration occurs when two or more teams work closely together to achieve a common goal. The teams involved in collaboration share responsibility for the outputs, or parts of the outputs, generated by their joint activities. The greater the number of shared outputs, the more overlap there is, and consequently, the higher the level of collaboration. When responsibility for all outputs is shared, the teams function as a single unit. Collaboration is the optimal mode of interaction when teams need to generate new ideas, explore new practices or technologies, or solve complex problems that require joint effort.

Collaboration enhances internal communication between team members, which fosters exploration and discovery but can reduce overall productivity. Therefore, when two or more teams collaborate, the goal they work on should be of high value to justify the productivity loss associated with collaboration. The reduction in total output should be justified by the outcome. Generally, this mode of interaction should be planned with clear, shared objectives and practiced for a limited time, aimed at achieving these objectives. A limited number of teams that need to collaborate indefinitely usually indicates that their scope and responsibilities need to be redefined. Conversely, a large number of teams collaborating indefinitely often suggests that establishing a federated team might be useful. This federated team would include only the individuals from each team who need to collaborate, typically for shared decision-making or coordination.

XaaS

XaaS occurs when one team interacts with another by providing a service, such as an API, a tool, or a product, that can be consumed with minimal interaction between the teams. This mode of interaction is the opposite of collaboration. If the service that's provided is well-designed and stable, it decouples the teams and allows each to focus on their own objectives, thereby maximizing delivery flow. However, if the service isn't well-designed, it can slow down the flow for the teams consuming the service. Additionally, because the service needs to be reasonably stable over time, opportunities for innovation at the boundaries between the teams can be limited.

Typically, the team providing the service and the team(s) consuming it might initially work together collaboratively to explore and design the service. Once the service has been defined and implemented, the interaction can shift to an XaaS model to streamline ongoing operations.

When establishing an XaaS interaction between two teams, an upstream-downstream relationship is created. The team providing the service (upstream) influences the team consuming it (downstream), but not vice versa. It's essential to clarify how this relationship is governed, including who decides on the service's offerings and its evolution. The context mapping patterns that were discussed in *Chapter 5, Designing and Implementing Data Products*, can be useful for this purpose. Specifically, the following four scenarios can occur:

- **Open host service**: The team providing the service determines its functionalities and how it will evolve. The consuming team or teams must adapt to these changes.

- **Customer/supplier**: The consuming team (customer) specifies the necessary functionalities and their evolution. The providing team (supplier) is responsible for implementing these requirements.

- **Partnership**: The team providing the service and the consuming team(s) define the functionalities and their evolution collaboratively.

- **Published language**: A third team, typically a federated team, defines the service's functionalities and evolution. Both the implementing team and the consuming team(s) adapt to these predefined specifications.

Facilitation

Facilitation occurs when one team helps another team solve a specific problem. The goal of this assistance is to make the supported team self-sufficient in addressing the issue. Facilitation aims to enhance the capabilities of the assisted team rather than taking over the problem-solving process. It is the preferred mode of interaction for enabling teams and often for federated teams as well. Effective facilitation requires expertise in the relevant areas and the ability to transfer knowledge.

A team may interact with multiple other teams simultaneously, using various interaction modes. Both team structures and interaction methods can evolve based on needs. Generally, interaction modes adjust more quickly to operational demands, while team structures remain more stable to preserve the achieved teams' cohesion.

The following figure illustrates how we will graphically represent the different modes of interaction when describing the team structure (*Figure 9.5*):

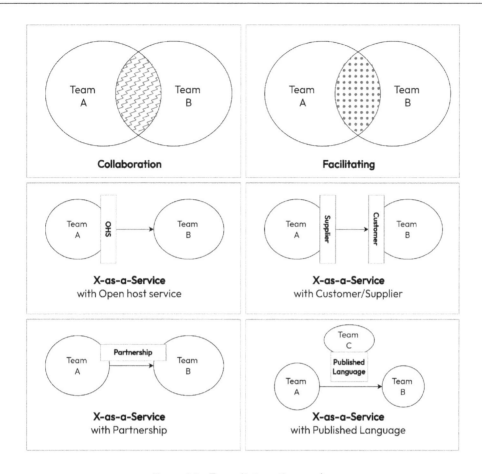

Figure 9.5 – Teams' interaction modes

Let's learn how to map the organizational architecture. We'll start by describing each team that composes it.

Mapping the organizational architecture

Each team has a leader who acts as the link between the team structure and the organizational chart within the organizational map. Although teams may interact horizontally across the chart, each team belongs to a specific organizational entity (for example, business unit, division, and department) based on its leader. A team's operations are influenced by its structure and the horizontal relationships with other teams. However, the methods for setting objectives and obtaining the necessary resources are determined by the team's position in the organizational chart.

Thus, the organizational chart and team structure must be aligned to achieve the organization's strategic goals. While the organizational chart and operating model ensure strategic alignment, this alignment is meaningless if the team structure can't execute the strategy efficiently. Conversely, an optimized team structure, which ensures excellent execution, loses its effectiveness without strategic alignment. Therefore, the organizational structure should be designed holistically, with equal emphasis on all its components.

Although the organizational chart and team structure are closely connected, they have separate life cycles. The organizational chart can change without affecting the team structure and vice versa. Typically, the team structure changes more frequently than the organizational chart. This doesn't mean that existing teams are constantly disbanded and reassembled; teams should remain stable to capitalize on the cohesion that's been achieved. However, the relationships between teams may shift more often, and there may be a need to create new teams to address emerging needs.

It's crucial to give the same level of attention to mapping the team structure as is typically given to defining and representing the organizational chart. Each team should have a clear description of its activities and interactions with other teams. This description, often called the **team API**, can be formalized using a **Suppliers, Inputs, Process, Outputs, and Customers (SIPOC)** diagram or any other type of structured document. The following figure provides an example of a document template that can be used to represent a team's APIs (*Figure 9.6*):

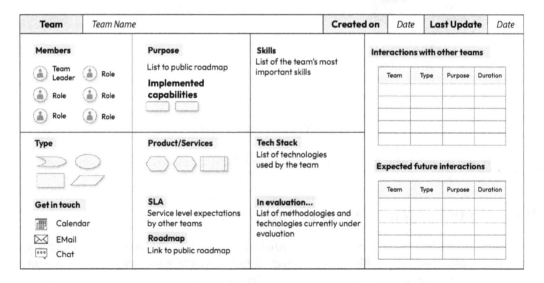

Figure 9.6 – Team API

Within the organizational architecture, each team has specific operational or management responsibilities. The allocation of these responsibilities across the team structure can lead to either more centralized and hierarchical business architectures or more decentralized and fractal business architectures. In the remainder of this chapter, the more decentralized model will be used as a reference for designing the architecture of a data management function oriented toward managing data as a product since it's more flexible and adaptable to change. Next, we'll examine how high-level responsibilities are distributed across teams in a distributed and fractal organizational model.

Delving into the fractal organization

Organizational architecture is a component of business architecture. Initially, business architecture outlines the value streams that support the business model and the business capabilities needed to realize them. Here, organizational architecture defines how these capabilities are instantiated and governed within the organization. Business capabilities are generally grouped by function and sub-function, while their instances are grouped based on the organizational architecture's structure.

The organizational architecture is influenced by the logical structure of the capabilities grouped by function and sub-function, but it doesn't necessarily coincide with it.

As discussed in *Chapter 4, Identifying Data Products and Prioritizing Developments*, the same capability can be instantiated by multiple organizational units belonging to different parts of the organization. Replicating the implementation of the same capability across multiple units, rather than concentrating it in a single unit, can be driven by operational efficiency reasons, such as specializing a particular capability by geography, product, or customer segment. Similarly, instantiating the same capability across multiple units can help decentralize specific control and management activities.

In *Chapter 3, Data Product-Centered Architectures*, we introduced the **viable system model** (**VSM**) to describe the key capabilities of a data management function focused on treating data as a product. According to the VSM, any organization aiming to evolve in a changing environment must be structured as a composition of subsystems, which, in turn, can be composed of smaller subsystems recursively. Each subsystem in the VSM comprises a subsystem that implements operational capabilities and systems that implement management capabilities. These operational capabilities are specific to the system, while the management capabilities remain the same across all systems, meaning they are multiple instances of the same management capabilities.

The VSM results in a fractal organizational structure where each nested subsystem mirrors the structure of its parent system. This fractal structure enables the distribution of strategic planning and control capabilities throughout the organization. Organizations that adopt this distributed model of strategic planning and control tend to be more agile and better equipped to adapt quickly to their environment compared to organizations that centralize strategy planning activities, keeping them separate from operational activities (*Figure 9.7*):

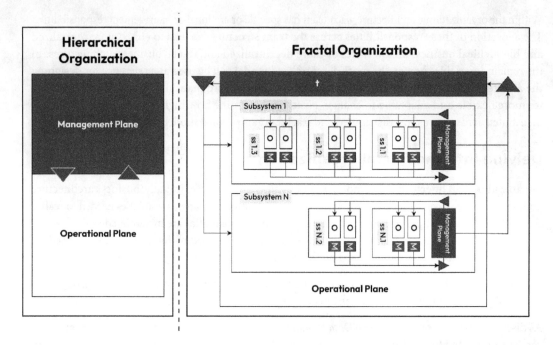

Figure 9.7 – Hierarchical versus fractal organization

In the following sections, we'll explore how to structure teams within the data management function so that we can implement both the operational and management capabilities needed to manage data as products. We'll use the VSM as a structural blueprint and the Team Topologies framework to define team types and their interactions.

Defining operational teams

A data management function oriented toward managing data as a product must have four core operational capabilities: data product development, XOps platform engineering, governance policy making, and data transformation enabling. This section will explore how to structure teams so that they can implement each of these capabilities optimally according to the organizational context.

Data product teams

Teams that implement the **data product development** capability within the data management function are stream-aligned teams that can develop and manage data products autonomously throughout their entire life cycle. They're aligned with business domains rather than layers of the technology stack. Each data product team is typically associated with one or more instances of business capability and is responsible for supporting the teams that implement these capabilities within the organization by developing and maintaining the necessary data products.

In small organizations, a single data product team may be associated with capability instances across multiple domains. In medium-sized organizations, a data product team is generally associated with all capability instances within a specific domain, with each domain having a dedicated data product team. In large organizations, even within the same domain, capability instances can be distributed among multiple data product teams (*Figure 9.8*):

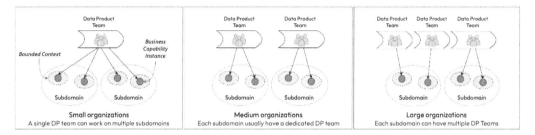

Figure 9.8 – Data product teams

When distributing business capability instances among data product teams, it's essential to consider the team's cognitive capacity. Ideally, a data product team should handle capability instances from the same domain or closely related domains. It's also important not to overload a data product team with more capability instances than it can manage effectively.

Data product teams interact with each other and with other business teams primarily in XaaS mode through the data products they develop. By definition, a data product is designed to make a data asset easy to use and compose with other data assets. Consequently, a data product should require minimal interaction between the consumer and the data product team that develops it for effective use.

Although a data product is developed to support the needs of downstream consumers, it's preferable not to establish a customer-supplier relationship between the product team and its consumers. This avoids tailoring the data product too closely to a specific consumer's needs, which can reduce its reusability. Generally, it's better to use the Open Host Service pattern to ensure greater reusability of the data product, and potentially complement it with the Published Language pattern to facilitate easier composability with other data products.

Product teams can also collaborate for short periods, primarily due to project needs or reorganization reasons. In the first case, one or more product teams temporarily work together to support a common business case by developing new data products. For example, in Lux, the e-commerce team and the central IT team collaborate on the real-time inventory business case, which requires both teams to implement new data products (that is, Digital Sales DP and Store Sales DP). Typically, in these situations, collaboration is strong during the initial phases of identifying and designing the products, then diminishes during the later stages, ending once the necessary implementations to support the new business case are completed. In the second case, collaboration occurs when a data product team becomes too large and needs to be split. Usually, after this split, the two new data product teams may continue to collaborate until they achieve full autonomy.

Each data product team must have a **data product owner**, sometimes also referred to as a data product manager, who acts as the primary interface between the team and the external ecosystem. The data product owner is responsible for defining the product roadmap so that it's in line with the organization's operating model and ensuring alignment with the business strategy objectives. Depending on their needs, organizations can define additional specific roles within each data product team. For example, it's common to have a technical leader role alongside the data product owner. The data product owner is an expert in the business domains to which the team is aligned and is responsible for understanding and prioritizing what needs to be done to support them. Generally, the data product owner may not have strong technical skills. Therefore, the technical leader acts as the main point of reference for the team and often for others outside the team, guiding how things should be done from a technical perspective. The technical leader is responsible for the quality and reliability of the products that are developed by the team.

Apart from the data product owner and other centrally defined roles, each team is free to organize its working methods as it prefers. Generally, data product teams work in an agile manner (for example, Scrum, Kanban, and so on). However, some data product teams may choose to adopt more traditional work practices, such as Lean, Waterfall, Six Sigma, and others. Typically, the more uncertain and complex the nature of the products a team needs to develop, the greater the benefit of adopting agile methodologies. Conversely, the clearer, more linear, and easily definable the products are upfront, the more advantageous traditional approaches can be. This might be the case with certain types of source-aligned products, for example.

The product owner is the link between the data product team and the rest of the organizational structure. The team is part of the organizational unit to which the product owner belongs. If the data management function is entirely centralized within the IT department, the product owners will be IT personnel reporting hierarchically to the head of this department (for example, CIO, CDO, and so on). Conversely, if the data management function distributes the ownership of data products across business domains, the product owners will be individuals from the specific domain to which the data product team is associated. In these cases, it's sometimes possible that while the product owner reports to the business domain, the more technical team members continue to report to IT. This situation can be useful during a transition phase when the business domain doesn't have the technical resources to create an autonomous, cross-functional team yet. However, it creates a matrix organizational structure where the technical team members have two supervisors: the product owner and someone within the IT department. This setup isn't optimal as matrix structures often lead to friction and organizational issues. It's preferable for all team members to eventually belong to the same organizational structure, either the business domain or the IT department, depending on whether a more centralized or decentralized data ownership approach is chosen.

So far, we've explored how data product teams implement the operational capability of data product development within the data management function. Now, let's examine how the teams that support other operational capabilities within the data management function are structured to support the stream-aligned data product teams. We'll start with the XOps platform engineering capability.

Platform teams

The teams that implement the **XOps platform engineering capability** within the data management function are naturally platform teams. Initially, there's generally a single platform team. Medium to small organizations with only one data product team might be tempted to also assign the responsibility of developing the XOps platform to this team. If this is unavoidable due to limited resources, it's still crucial to always keep the budgets for the two types of activities – platform development and product development – clearly separate. Otherwise, there's a risk that the urgency to release new data products will cannibalize the time and budget needed for the efficiency and rationalization activities brought by the platform.

As adoption scales and the platform's complexity increases, the team managing it may need to split into multiple teams. Typically, each team specializes in different layers of the platform. For example, one or more teams might focus on developing the utility plane, others on the data product control plane, and others on the data product marketplace. With growing complexity, it may also be necessary to create complicated-subsystem teams specialized in cross-cutting concerns such as security, FinOps, or data ingestion from specific legacy systems. Ingestion teams are particularly common within the platform team structure. Although they develop true source-aligned data products, they're generally associated with the XOps platform engineering capability since these products don't involve business logic. These data products aim to manage sourcing by masking the complexities of the source legacy system to downstream data product teams.

Finally, in extremely complex organizations, platform teams might be organized into a specific sub-function within the data management function, with its own dedicated management system. This system then coordinates with the parent data management function's management system, as shown in *Figure 9.9*:

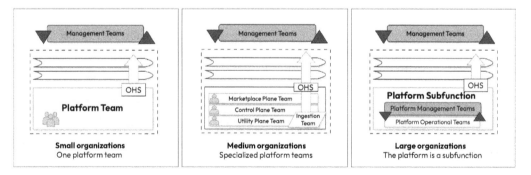

Figure 9.9 – XOps platform teams

Primarily, platform teams interact with product teams by providing features or tools in XaaS mode. For the design of specific platform functionalities, they may collaborate with product teams for limited periods. This often occurs when a feature that's already being implemented independently by one or more product teams needs to be integrated into the platform.

Each platform team has a **platform team leader** who typically reports to a manager within the IT department. The XOps platform engineering capability isn't distributed across the organization. Given its mission to unify technological elements, it's generally centralized within the IT department. When there are multiple platform teams, it's beneficial to establish a federated team that includes the leaders of each team to ensure coordinated platform development.

Governance teams

The teams that implement the **governance policy-making capability** within the data management function can be dedicated enabling teams or federated committee-type teams. In some cases, they can also be hybrid teams, where some members work continuously for the team while others are only partially involved. The former are responsible for the team's operations, while the latter participate in activities to support the decision-making process, as defined by the operating model.

A federated or hybrid structure is generally preferred to ensure the involvement of representatives from all teams that will be impacted by the policies. These teams typically include representatives from the data product teams that will comply with the policies, platform teams that will enforce them, and enabling teams that will facilitate their application.

Even for committees or hybrid teams, it's important to keep the team size limited (that is, a maximum of 15 people, ideally no more than nine). To meet this operational constraint while ensuring the participation of all relevant stakeholders, it's useful to break these teams down according to the type of policies they're responsible for defining (for example, policies related to technologies, data quality, APIs, communication protocols, and so on). In *Chapter 11, Building an AI-Ready Information Architecture*, we'll explore the roles and functions of the governance team that are dedicated to defining the enterprise conceptual model that's necessary to ensure semantic interoperability among the developed data products.

In complex organizations, governance teams can also be organized hierarchically. For instance, one or more teams can establish global policies for the entire data management function, while local teams can define policies specific to individual domains. (*Figure 9.10*):

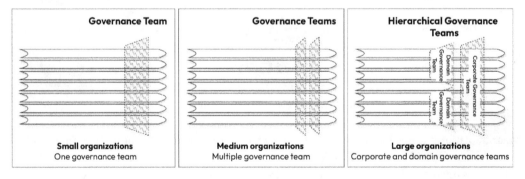

Figure 9.10 – Governance teams

Within federated governance teams, various teams collaborate through their representatives to make joint decisions on policies for the data ecosystem. Externally, governance teams interact with other operational teams primarily in XaaS mode through the governance policies they define. These policies should be managed as products that are easy to understand and apply without requiring direct interactions with the governance team itself.

Each governance team must have a **governance team leader** who represents the team externally. The team leader is responsible for the team's proper operation but not necessarily for the decisions made. Since the team is federated, decisions are made collectively according to the operating model. Federated governance teams don't belong to any specific department within the organization but operate cross-functionally. Generally, the team leader of governance teams reports to the IT department.

Enabling team

The teams that implement the **data transformation enabling capability** are enabling teams. Initially, these teams may not exist, and the responsibility for this capability may fall to platform and governance teams. However, as the adoption of the new paradigm grows, dedicated teams focused on enabling activities become necessary. In small and medium-sized organizations, a single enabling team is often sufficient. In larger organizations, multiple enabling teams are often needed to better scale support and change management activities. Having multiple enabling teams with the same skills and responsibilities allows for parallel support of more product teams, preventing the data transformation enabling capability from becoming a bottleneck. Additionally, multiple enabling teams can be created not only for scaling purposes but also to separate specific competencies. In this case, vertical enabling teams are formed for specific facilitation areas such as onboarding, compliance, and quality (*Figure 9.11*):

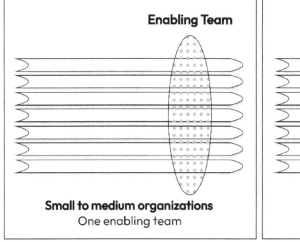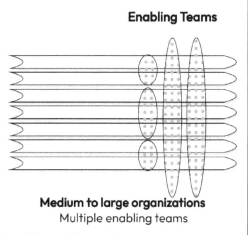

Figure 9.11 – Enabling teams

The primary mode of interaction for enabling teams is facilitation. These teams work with both data product teams and governance and platform teams. Their role is crucial not only in enhancing the operations of data product teams but also in gathering and sharing feedback with governance and platform teams, ensuring that their activities are closely aligned with the actual needs of the data product teams.

Each enabling team must have an **enabling team leader**, who is responsible for the team's facilitation activities. Enabling teams are generally part of the IT department, with their team leaders reporting directly to a manager within the department.

With this, we've explored the types and structures of all teams responsible for implementing the operational capabilities of a data management function oriented toward managing data as a product. In the next section, we'll explore the types and structures of teams that are responsible for management capabilities.

Defining management teams

A data management function oriented toward managing data as a product must possess the managerial capability to define its identity, derive the optimal data strategy, and plan and control its execution. In this section, we'll explore how to structure the teams responsible for implementing these managerial capabilities effectively.

Data strategy committee

The **data strategy committee** is responsible for defining the identity and high-level strategy of the data management function. Within the VSM, it's the team that implements the functions of System 5 (Identity) and System 4 (Strategy).

It's generally a federated team that's composed of the leaders of the data management function. In centralized architectures, the leaders of the data management function are the same as the leaders of the IT function, who are responsible for data management. In decentralized architectures, the data strategy committee consists of the top figures responsible for the data function in each domain.

The data strategy committee coordinates with the teams responsible for defining the organization's overall identity (leadership team) and strategy (top management team) to ensure alignment between the business strategy and the data strategy.

At the operational level, the strategy for the data management function is defined according to the adopted operational model. In the case of EDGE, the data strategy committee is responsible for establishing the vision for the evolution of the data management function and translating it into measurable goals through measures of success and fitness functions. This process is iterative, occurring in semi-annual or annual cycles. During each cycle, the vision and goals are reviewed and updated based on insights gained from the previous iteration and any external changes in the organization's overall business strategy.

In each iteration, the data strategy committee is also generally responsible for negotiating the financial resources needed to support the execution of the data strategy for the next cycle with the rest of the organization and allocating the obtained resources to the various identified goals, thereby defining data strategy priorities.

Data portfolio management committee

The **data portfolio management committee** is responsible for strategic planning and controlling the execution activities. Within the VSM, it's the team that implements the functions of System 3 (Control) and System 3* (Monitor and Audit).

It's generally a federated team that's composed of the leaders of the various teams or sub-functions within the data management function.

The data portfolio management committee coordinates with the teams responsible for strategic planning and execution across the organization (top management team and demand team). It receives priorities for the initiatives the organization intends to pursue from these teams. In turn, it reports on the results of the data management initiatives and how they're contributing to the broader goals of the organization's strategic plan.

At the operational level, the strategic planning of the data management function is defined according to the adopted operating model. In the case of EDGE, the data portfolio management committee is responsible for translating the vision and goals defined by the data strategy committee into operational initiatives. This activity is conducted iteratively, in quarterly or semi-annual cycles. In each cycle, the initiatives in the portfolio are reviewed and updated appropriately based on the results obtained and any potential changes made to the strategy goals by the data strategy committee.

In each iteration, the data portfolio management committee is typically responsible for allocating the resources that have been assigned to each goal by the data strategy committee to the initiatives identified to achieve those goals and the activities defined to implement them. However, the specific activities required to implement each initiative aren't determined directly by the data portfolio management committee. Instead, following the subsidiarity principle characteristic of EDGE, this responsibility is delegated to the operational teams that will be tasked with their execution.

To ensure optimal resource management and the harmonious evolution of the initiative portfolio, all requests from the organization to the data management function must go through the data portfolio management committee. Directly exposing operational teams to the organization's demands risks undermining the coherence of the data management function's strategic plan.

Operations management committee

The **operations management committee** is responsible for coordinating execution activities. Within the VSM, it's the team that implements the functions of System 4 (Coordination).

The operations management committee can either be a federated team composed of the leaders of various operational teams or a dedicated team focused on managing projects that involve multiple operational teams, such as a project management office. In organizations of medium complexity, each initiative typically requires the coordinated execution of activities to be handled by different teams. Even with an optimal team topology, interdependencies are inevitable. In such contexts, having a federated or dedicated team specifically responsible for coordinating activities is crucial. This team ensures that tasks are completed on time and that each team has the necessary resources to do so.

Putting it all together

The data management function consists of a network of teams working together to make the organization's data assets accessible and easily usable to support multiple business cases in line with the business strategy. The structure of these teams, their responsibilities, the ways they interact, and the operational model used to make decisions and allocate resources define their architecture. The following figure shows the overall architecture of a data management function oriented toward managing data as a product, as described so far (*Figure 9.12*):

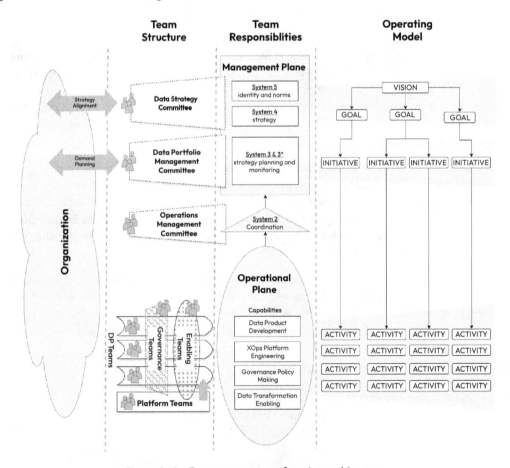

Figure 9.12 – Data management function architecture

The types of teams involved and their interactions are described using the Team Topologies framework. The distribution of responsibilities among these teams is based on the VSM, while the operational model is centered on the lean value tree defined by the EDGE methodology. Team Topologies, the VSM, and EDGE work synergistically and convergently to define a modular, adaptive architecture optimized for fast delivery flow, which is ideal for managing data as products.

In the next section, we'll see how this logical architecture integrates into the organizational structure, leading to more or less decentralized data ownership across business domains.

Evaluating decentralization strategies

Unlike the data mesh paradigm, the paradigm centered on managing data as a product doesn't require the decentralization of data ownership to business domains. Decentralization is an option that's enabled by modularizing the data management solution through data products, but it's not a necessary outcome. In this section, we'll explore when it's advantageous to decentralize ownership, which data to decentralize, and the strategies for successfully implementing decentralization.

Understanding when to decentralize

By managing data as a product, it's possible to create modular data management solutions that are easier to maintain and evolve compared to traditional monolithic solutions. Managing data as a product is a practice that can be adopted by any organization, regardless of its size or business context.

Modularizing data management solutions through data products allows for the distribution of data ownership. Rather than having a single team manage the entire solution, multiple teams can be assigned various roles and responsibilities. Specifically, stream-aligned teams responsible for developing data products can be shifted from the IT department to the relevant business domains, effectively decentralizing data ownership.

Decentralizing data ownership to business domains, a key principle of the data mesh approach, enables a more scalable and responsive data management function. However, this decentralization introduces significant complexity both during the adoption process and in ongoing operations. Therefore, when beginning to adopt the practice of managing data as a product, it's essential to carefully assess whether decentralizing data ownership to business domains is advantageous while weighing the associated costs and benefits.

Generally, it isn't advisable to decentralize data ownership in the following instances:

- The organization, due to its size or life cycle stage, has reasonable scaling needs for the data management function that can be perfectly addressed just with a centralized ownership model

- There's sufficient sponsorship to transform the data management function within the IT department, but not enough to drive significant organizational changes beyond IT

- Organizational maturity in data management is low, and adding decentralization to the adoption of the data-as-a-product paradigm would make the risks of the initiative too high

- The likelihood of obtaining the necessary buy-in from business domains for the initiative is low, even with strong sponsorship from the top

- The organization faces resource or time constraints that are incompatible with such a complex organizational transformation

Even when the benefits of decentralization outweigh the associated risks and costs, it can still make sense to evaluate whether to start with a decentralized model immediately or adopt the data-as-a-product approach first and address decentralization only after it has gained traction. In many cases, a more cautious and gradual approach is advisable.

Understanding what to decentralize

Data ownership can be either fully centralized within the IT department or completely decentralized across business domains. However, these extremes represent only the two ends of the spectrum. Between them, there are intermediate decentralization solutions that can be tailored to fit the organization's needs and its data management maturity. For instance, data can be categorized – and by extension, the data products that provide access to this data – in various ways. Based on this classification, decisions can be made on which types of data should have decentralized ownership in business domains and which should remain centralized within the IT team.

A useful approach for deciding whether and when to decentralize data ownership is to categorize data based on its level of enrichment. On one side, there's raw data, which is closer to the sources and is typically managed by data products that are source-aligned. On the other side, there's transformed data, which is optimized to support specific business cases, closer to the consumers, and generally managed by data products that are consumer-aligned. Based on this classification, intermediate decentralization strategies can be adopted. For instance, ownership might be decentralized only for raw data and its corresponding source-aligned data products, or exclusively for transformed data and its consumer-aligned data products (*Figure 9.13*):

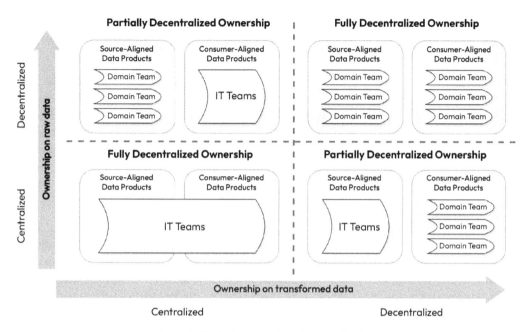

Figure 9.13 – Data ownership decentralization

In the process of decentralizing data ownership to business domains, partial decentralization can serve as either an optimal end goal or a transitional step toward complete decentralization, depending on the organization's size and needs. Now, let's look at how to structure this path toward data ownership decentralization.

Moving toward decentralization

Data can be managed as products without necessarily decentralizing ownership to business domains. However, the reverse isn't true. To sustainably decentralize data ownership to domains without creating data silos that can't interoperate, it's essential to manage data as a product. Therefore, data ownership decentralization should be done either alongside the adoption of the data-as-a-product paradigm or at a later stage. Typically, if decentralization is pursued, it's wise to begin by adopting the managing data-as-a-product paradigm while minimizing impacts on the organizational structure required by ownership decentralization during the early adoption stages. This way, attention can be focused on establishing core practices, the operating model, initial platform functionalities, and the first lighthouse projects, all of which are foundational for decentralizing data ownership toward domains.

Once the initial bootstrap phase is complete, it's possible to begin integrating data ownership decentralization into the adoption of the new paradigm. It's essential, even during the bootstrap phase, to design practices, the operating model, and the core platform to support the future decentralization of ownership. However, it's important not to start decentralizing ownership too early, especially when there's significant uncertainty about practices, the operating model is still being refined, and the platform lacks the essential services needed to support data product development. Doing so might introduce excessive complexity for the domains responsible for managing their own data, potentially leading to low buy-in or even outright rejection of the new paradigm.

Between the end of the bootstrap phase and the beginning of the scaling phase is generally the best time to start the incremental process of decentralizing data ownership. It's crucial to carefully choose the order in which domains are transitioned to decentralized ownership. Typically, it's advisable to start with domains that already make extensive use of data and have the internal resources to form cross-functional teams for managing data products. It's also important to assess the willingness of a domain and its key figures to take on the responsibility for the data they manage. Even with strong top-down sponsorship, it's always preferable to start with domains that are open to change and willing to collaborate, rather than those that are resistant to change and more likely to oppose the new approach.

Once the domains to start with have been identified, they should undergo operational onboarding, supported by enabling teams until they become self-sufficient in managing their data products. Depending on the capacity of the enabling teams, it's possible to onboard multiple domains in parallel. However, it's crucial to avoid attempting to decentralize ownership to too many domains simultaneously without the necessary resources to support them. Overextending can lead to dissatisfaction, decrease buy-in, and ultimately risk the initiative's success. In this case, success should be measured not by the number of domains to which ownership has been decentralized, but by the number of domains that are capable of autonomously managing the ownership they've been given.

Ownership can be shifted to a domain for specific types of data only. As discussed in the previous section, many organizations choose to decentralize ownership of only raw data to business domains, while others opt to decentralize only ownership of transformed data.

Decentralizing ownership solely for transformed data is generally the most popular approach. The central IT department handles data sourcing from legacy systems, typically through ingestion teams that create source-aligned data products. Domain product teams, on the other hand, develop consumer-aligned products to support the organization's business cases. This approach is commonly adopted because creating source-aligned products generally requires significant technical expertise, which is provided by the IT department. Conversely, developing consumer-aligned products demands a deep understanding of business requirements, making it more suitable for domain teams to handle directly.

Decentralizing ownership solely for raw data is common in product-centric digital companies. In these cases, the source systems aren't legacy systems that are managed by IT for the domains but are applications that are directly developed by the domains for their external and internal clients. Therefore, the domains are better equipped to develop the source-aligned products associated with these systems. Ownership of enriched data, on the other hand, remains centralized within the IT department or another specific domain and focuses on creating new products generally aimed at the external monetization of data collected from the applications managed by the domains.

The decentralization patterns outlined here serve as general models that guide decisions about what and when to decentralize. However, in practice, achieving a clear-cut approach to decentralizing ownership can be challenging as business domains within an organization can vary widely, as can the types of data they handle. Given a general decentralization strategy, it's normal to have various exceptions that account for these differences. For example, in a partial decentralization strategy, it might be practical to grant large domains full accountability over their data, while smaller domains may fully delegate the management of their data to the central IT department.

Summary

In this chapter, we learned how to define the organizational structure needed to implement a data management function capable of managing data as products. In particular, we focused on the modular unit of this architecture: the teams.

First, we introduced the Team Topologies framework, outlining its four types of teams and the three types of interactions between them.

Then, we applied these key concepts to define the structure, types, and interactions of the teams necessary for implementing the organizational architecture of the data management function. Here, we examined the data product teams, platform teams, governance teams, and enabling teams responsible for implementing the operational capabilities of the data management function. Then, we analyzed the data strategy, data portfolio management, and operations management committees, all of which are responsible for implementing the management capabilities of the data management function.

Finally, we explored how and when to complement the data-as-a-product approach with the decentralization of data ownership toward business domains. We proposed various decentralization methods and recommended a critical and incremental approach to implementing decentralization.

In the next chapter, we'll learn how to manage data modeling within a modular and distributed architecture.

Further reading

For more information on the topics that were covered in this chapter, please check out the following resources:

- *Organization Mapping*, by TOGAF (2022). Available at `https://pubs.opengroup.org/togaf-standard/business-architecture/organization-mapping.html`.

- *Team Topologies: Organizing Business and Technology Teams for Fast Flow*, by M. Skelton and M. Pais (2019). Available at `https://a.co/d/3hDLLbB`.

- *Let's Unfix Team Topologies*, by J. Appelo. Available at `https://unfix.com/blog/lets-unfix-team-topologies`.

- *Systems Thinking: Combining Team Topologies with Context Maps*, by M. Plod. Available at `https://www.youtube.com/live/xbH2rxXsaI0?si=yhJaNwetgGdrX6Ko`.

- *Team API template*, by M. Skelton and M. Pais. Available at `https://github.com/TeamTopologies/Team-API-template`.

- The Fractal Organization, by Patrick Hoverstadt (2009). Available at `https://www.amazon.it/Fractal-Organization-Creating-Sustainable-Organizations/dp/0470060565`.

- *Data Mesh Applied*, by Sven Balnojan (2019). Towards Data Science. Available at `https://towardsdatascience.com/data-mesh-applied-21bed87876f2`.

Get This Book's PDF Version and Exclusive Extras

Scan the QR code (or go to `packtpub.com/unlock`). Search for this book by name, confirm the edition, and then follow the steps on the page.

Note: Keep your invoice handy. Purchases made directly from Packt don't require one.

10

Distributed Data Modeling

In the previous chapter, we explored the design of an organizational architecture that aligns with both the business strategy and the surrounding context to enable the adoption of the data-as-a-product paradigm effectively.

This chapter delves into the heart of data modeling within a modular, and often decentralized, architecture that revolves around data products. We begin by defining what a data model is at its core, identifying its components, and discussing why, in a data-product-centric architecture, intentional and explicit data modeling is essential. We will unpack the modeling process itself, breaking down its stages and key deliverables.

From there, we'll explore various physical data modeling techniques, evaluating their strengths and limitations, and outlining how each can be applied within the distributed nature of a modern data-product-centric architecture. Finally, we will focus on conceptual data modeling, examining how these high-level abstractions can help in the design and usage of data products. We'll discuss the methods for crafting conceptual models, looking closely at the activities required to design and evolve them in harmony with a decentralized ecosystem of data products.

This chapter will cover the following main topics:

- Introducing data modeling
- Exploring distributed **physical modeling**
- Exploring distributed **conceptual modeling**

Introducing data modeling

Before delving into the various data modeling techniques applicable in a distributed environment, it is crucial to first understand what a data model truly is—how it is constructed and formally represented to enable seamless sharing and interoperability.

What is a data model?

In *Chapter 4, Identifying Data Products and Prioritizing Developments*, we discussed how, in **domain-driven design** (DDD), modeling the domain where the problem resides—what we call the **problem space**—is fundamental to crafting an effective solution. But let's unpack what a model really is. At its core, a **model** is an abstraction, a tool that enables us to navigate the intricacies of a domain by stripping away the noise and focusing on the elements that are truly critical to solving the problem at hand.

A model is a simplified representation of a thing or phenomenon that intentionally emphasizes certain aspects while ignoring others. Abstraction with a specific use in mind. (Rebecca Wirfs-Brock)

A single domain can be modeled in different ways, depending on the problem you're aiming to solve. The quality of a good model isn't determined by how accurately it mirrors the reality it describes, but by how effectively it addresses the problem it was designed to solve. The best models are those that offer a solution with the fewest possible details—among all the models that can solve a problem, the simplest ones are the best. In general, a model serves as the foundation for building a solution to a problem. Those who benefit from the solution don't need to fully understand the model that led to its creation. However, when it comes to data, the model is an integral part of the solution itself. It's through the model that data is structured and transformed into information that can be more or less useful to consumers, depending on the problem they're trying to solve.

The usability and reusability of a data model are often at odds with each other. Data models that are too closely tied to a specific problem lack the generality needed to be reused for solving other problems. Conversely, models that are too general tend to be more complex, making them harder to use compared to simpler models. It falls to the modeler to strike an optimal balance between usability and reusability of the data model. In this chapter, we'll explore how to achieve that balance.

Implicit and explicit models

Every time we set out to solve a problem, we create a mental model of the reality in which the problem exists and use that model to craft a solution. A mental model is an **implicit model**—one that lives in the mind of its creator and is partially, and only implicitly, reflected in the solution produced. It can be inferred from the solution, but it's not easy to separate it from the solution or to reason about it, even for the person who originally conceived it.

Designing a solution for a complex problem often requires the collaboration of individuals with diverse skills and backgrounds. Without a shared model of the problem space, each person involved operates with their own implicit model of the problem, shaped by both the information they've received and their prior experiences. Communication among team members then becomes a constant exercise in translating from one implicit model to another.

For instance, an analyst defining requirements knows what they intend to communicate, but they cannot fully grasp how the recipient, such as the development team, interprets that information or how they adjust their mental models of the problem in light of the new information. This turns the design process into a game of telephone, where useful information is continually exchanged but each participant maintains their personal model of the problem. In the absence of a shared, explicit model, much of the information is lost in the ongoing translation between implicit models. This leads to misunderstandings and misalignments regarding the solution being developed.

Therefore, it is crucial for all involved in the design process to collaborate in creating a shared, **explicit model**, formalized in a language that everyone understands. Each piece of relevant information is then represented within this shared model. The explicit model becomes the focal point of the design process, allowing all participants to confirm and document their common understanding of both the problem and the solution being built.

The lack of a model is still a model. It's just a crappy model. (Joe Reis)

For solutions developed as products, the presence of a shared model is even more crucial. Products have a life cycle and evolve over time. Thus, the value of the model does not end with the initial release of the solution but continues to underpin subsequent iterations and evolutions.

In the case of pure data products, explicit modeling is important not only for the development and evolution of the solution but also to enable potential consumers to understand and utilize the product effectively.

Let's now see how the modeling process can be structured and what its main outputs are.

Data modeling process

A **data model** is a composite and layered construct. It is *composite* because it integrates multiple models, each serving a specific purpose and using its own formalism to represent the reality being analyzed. It is *layered* because these models are interdependent and developed sequentially, with each model building on the previous one—either by enriching or specializing it—and laying the groundwork for the models that follow in subsequent stages.

A data model is typically divided into three distinct layers, each represented by a specific submodel: the conceptual layer, the logical layer, and the physical layer. Each submodel corresponds to one of the three primary stages of the data modeling process (*Figure 10.1*): conceptual modeling (alignment phase), logical modeling (refine phase), and physical modeling (design phase).

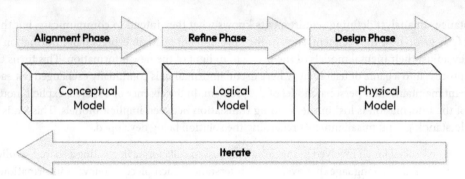

Figure 10.1 – Data modeling process and deliverables

The **alignment phase** centers on clearly modeling the **problem space**. In this phase, the goal is to identify all key concepts relevant to the problem, clarify their meanings, and map out their relationships. For example, in a project to improve customer satisfaction, concepts such as *customer feedback*, *response time*, and *service quality* would be defined and linked. All stakeholders involved in the project actively participate in this analysis. The resulting conceptual model then provides a foundation for the next modeling stages.

The **design phase**, on the other hand, focuses on explicitly modeling the **solution space**. Its objective is to operationalize the conceptual model by technically defining how the data related to the identified concepts will be structured to facilitate its use in relation to the problem at hand and the tools used for implementing its solution. Only the most technical roles, responsible for building the solution, are involved in this modeling phase. The **physical model** produced as the output serves as the foundation for subsequent implementations. If the conceptual model defines what needs to be done, the physical model generated in this phase defines how to do it.

The **refine phase**, situated between the alignment and design phases, focuses on refining the conceptual model, making it easier to translate into a physical model. The **logical model** produced as the output of this phase reflects the structure that the data will take on a physical level while remaining independent of the specific technological implementation. It can thus be seen as a transitional model, an intermediate step, to bridge the problem space and the solution space.

In the modeling process, the refine phase is essential, but creating a dedicated logical model as its output isn't always necessary. Everything produced in this phase can be represented either as an extension of the conceptual model or as an initial blueprint for the physical model. For this reason, in the rest of this chapter, we will focus only on the conceptual and physical models, considering the logical model as an extension of the conceptual model.

Data model representations

The physical data model is inherently explicit. It is impossible to present data to users without clearly defining the structure that contains and makes it physically accessible. The conceptual model, however, can be either implicit or explicit. Only an explicit model can be shared among all stakeholders, forming a common understanding of the problem necessary for intentional physical model design and for informed data usage. Emphasizing the importance of having an explicit and shared data model refers to the model in its entirety. Both its components—the conceptual and the physical—must be designed, represented, and shared with equal attention.

To make a model explicit, it's crucial to choose a representation method that is both consistent and familiar to everyone who will interact with it, whether during the design phase or in its later use. The chosen method should be detailed enough to capture the necessary information but also straightforward enough to be accessible to a wide range of people. We can categorize representation methods into two families:

- **Formal representations**: These are highly structured and adhere to precise rules. They clearly define the symbols, syntax, and semantics of the language used to build the model. The goal of a formal representation is to create models that are accurate and unambiguous, ideally free from any interpretative variability. Formal representations are also typically machine-readable. This allows them to be processed by software for tasks such as validation, editing, creating derived artifacts, or converting them into other types of formal representations.

- **Informal representations**: These are less structured and follow looser rules. They are therefore more flexible and adaptable to various contexts, but also more prone to generating ambiguity and interpretative variability. The purpose of an informal representation is to share high-level ideas in contexts where the focus is on communication and interaction with a broad audience, including non-technical stakeholders, rather than on detailed implementation specifics. Typically, informal representations are used to model problems with business stakeholders. Once a sufficiently detailed informal model has been created, technical team members can then independently convert it into a formal representation with minimal risk of interpretative distortions.

In the realm of data models, conceptual modeling is generally constructed through informal representations that are accessible to everyone, such as concept/relationship diagrams and business term glossaries. In contrast, physical modeling is developed by more technical individuals using formal representations, such as SQL DDL or JSON Schema, typically tailored to the technologies used for storing and accessing the data.

Event storming, which we introduced in *Chapter 4, Identifying Data Products and Prioritizing Developments*, is an example of an informal representation method for business process modeling. Once the event storming analysis of a process is complete, there is usually enough shared information to translate the model into a formal representation, such as one based on the **Business Process Model and Notation (BPMN)** standard.

As data consumers increasingly include applications that are not specifically programmed to handle particular types of data but can work with generic data (e.g., knowledge base systems, semantic layers, autonomous AI agents), it becomes more important to formalize the representation of the conceptual model to provide these applications with as much context as possible. We will discuss how to formalize the representation of the conceptual model in the *Exploring distributed conceptual modeling* section of this chapter. Now, let's explore how data modeling methodologies can be classified according to the type of data being modeled.

Operational and analytical data modeling methodologies

Most of the data managed by an organization is generated directly by the operational applications that handle its core processes. This data is then typically collected, integrated, and made available for the development of analytical applications. The data management challenges faced by operational applications are often quite different, and sometimes orthogonal, to those encountered by analytical applications. Over time, distinct technology stacks and modeling frameworks have been developed to manage data according to these two usage scenarios.

Operational data modeling focuses on efficiently handling individual transactions by ensuring data is quickly written and updated according to business rules. The main priority is to optimize data writing, not reading. In contrast, analytical data modeling integrates data from various systems to support broader business insights. The main priority is to optimize data reading for analysis, not writing or updating.

The distinction between data modeling techniques for operational versus analytical purposes only becomes evident at the physical model level. The conceptual model of the data remains unchanged because the meaning that the business assigns to the relevant concepts for describing a problem and its solution does not vary; it is independent of how the associated data will be managed and consumed. For example, the concept of a "sale" within a specific domain is uniquely defined. A sale is a sale, regardless of whether the data representing it is used for operational or analytical purposes. However, the way the data is used impacts the design of the physical model. Within a domain—or more precisely, using the terminology of DDD, within a bounded context—there is generally a single conceptual model from which multiple physical models can emerge, depending on the intended use of the data (operational or analytical) and the type of technology used to store and access it.

The focus of this book is on data that exists outside of applications—both operational ones that generate the data and analytical ones that consume it—and how this can be managed as a product. Specifically, pure data products are the means through which data, extracted from the operational applications that generated them, is made easily usable and composable for both analytical and operational applications that need it. Since data managed by a pure data product is exposed externally in a read-only format, the rest of this chapter will delve deeper into the modeling techniques defined within the analytical realm, while keeping in mind the following points:

- A pure data product is inherently *multimodal*. It can present data related to a specific conceptual model through various physical models, exposed via different output ports.

- A pure data product must support multiple business cases through the data it manages, not necessarily limited to analytical purposes but potentially operational as well. Typically, operational applications can use pure data products for data exchange and master data management.

- Despite exposing data in a read-only format, a pure data product should strive to ensure excellent performance along the data transformation pipeline. This is crucial for supporting operational consumers, but also increasingly important for analytical consumers who require data to be updated in near real time to intervene directly in the processes generating the data analyzed while they are still ongoing (**continuous intelligence**).

In the upcoming sections of this chapter, we'll explore the nuances of modeling data within a distributed architecture centered on data products, beginning with a deep dive into physical modeling.

Exploring distributed physical modeling

In this section, we'll dive into how to approach physical data modeling within an architecture centered on data products. Specifically, we'll explore how to adapt widely used analytical data modeling techniques to fit modular and distributed environments. We won't delve extensively into platform-specific modeling aspects (e.g., data lakes, data warehouses, data lakehouses, etc.), as the methodologies discussed here are broad enough to be easily applied across various major technology platforms.

Dimensional modeling

Dimensional modeling is a foundational technique for managing analytical data, first introduced in the early 2000s by Ralph Kimball and Margy Ross in their seminal work, *The Data Warehouse Toolkit*. This approach quickly became the gold standard for designing business intelligence and data warehousing systems, serving as a blueprint for countless implementations. Over time, the core principles of dimensional modeling have been adapted and extended by various other data modeling techniques to address specific design and operational needs.

The essence of dimensional modeling lies in its ability to simplify data analysis. It achieves this by structuring data around a limited set of key entities, interlinked through a minimal, yet clear set of relationships. This design principle ensures that data is organized in a way that is both intuitive and effective for in-depth analysis, paving the way for more actionable insights.

In a dimensional model, the core components are facts and dimensions. **Facts** represent the key business processes or specific transactions occurring within these processes. For instance, in a retail context, such as LuX, a product promotion, sale, or shipment could each be modeled as a fact. Each fact is a composite entity encompassing all the relevant measures needed for analyzing the transaction at hand. For a sales fact, this might include metrics such as units sold, unit price, total amount, applied discount, and margin. **Dimensions**, conversely, provide the contextual backdrop for these transactions, enriching the analysis of the measures associated with each fact. They offer the necessary framework to understand and interpret the data. In the case of a sales transaction, dimensions could include the customer making the purchase, the product bought, the timing of the transaction, the sales channel used, and so forth. This contextual layering enables a more nuanced and insightful analysis of the business processes represented by the facts.

A fact is linked only to the dimensions along which its measures are analyzed. Physically, this relationship is represented by a **fact table**, which has a one-to-many connection with the **dimensional tables** that provide context for the fact's measures. Each dimensional table has a surrogate key (ex. `CustomerID`, `ProductID`, `TimeID`, etc.), which is referenced by the fact table via a corresponding foreign key. The collection of these foreign keys in the fact table, pointing to the associated dimensional tables, forms the fact table's primary key. This structure is designed to ensure that the fact table integrates seamlessly with its dimensions, facilitating a coherent and insightful analysis of the measures. The following figure (*Figure 10.2*) illustrates an example of a dimensional model related to sales data.

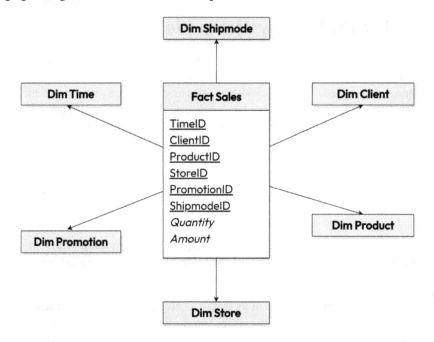

Figure 10.2 – Sales star schema

This highly denormalized model, with the fact table at its center and dimension tables radiating outward, is known as a **star schema**—one of the most prevalent physical structures for dimensional models. An alternative to the star schema is the **snowflake schema**, where dimension tables are partially normalized to streamline data loading processes and optimize storage efficiency. In either case, the fact table along with its dimension tables forms an n-dimensional space (i.e., hypercube) where measures can be easily analyzed from various perspectives by simply selecting specific coordinates (**slicing**) or groups of coordinates (**dicing**) on each dimension.

Using simple LEFT JOIN and GROUP BY clauses, or with the help of visual tools for **online analytical processing (OLAP)**, it is possible to analyze the units sold of a specific product across all stores in Italy, and then quickly switch to analyzing the units sold of all products in the accessories category within a particular store of interest, and so on.

Let's now examine how dimensional modeling fits within a centralized data architecture.

Centralized dimensional modeling

Dimensional modeling, as defined by Kimball and Ross, is framed within the broader data warehouse architecture they pioneered, now widely known as the **Kimball architecture**. This architecture is structured around two core layers: the staging area and the presentation layer. The **staging area** acts as a repository where raw data from various sources is cleansed, integrated, deduplicated, and historized. Once processed, the data is funneled into the **presentation layer**, where it is used to create data marts for consumer access (*Figure 10.3*).

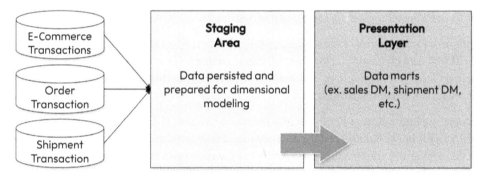

Figure 10.3 – Kimball data warehouse architecture

A **data mart** represents a set of consolidated, dimensionally modeled data focused on a specific business process or subject area, such as sales or shipments. These data marts align naturally with business domains, typically evolving under the guidance of domain teams in an independent and autonomous manner.

Data marts across different domains often share common dimensions and facts. For instance, the product dimension is leveraged by both the sales fact in the sales data mart and the shipments fact in the logistics data mart. Similarly, the sales fact might exist in multiple contexts, such as the data mart for in-store sales and the one focused on online sales.

To enable seamless interoperability between these data marts, it's essential that shared dimensions and facts are physically aligned in structure and populated based on consistent business logic. Facts and dimensions constructed in this way are referred to as **conformed facts** and **conformed dimensions**, respectively. This conformance ensures that different business domains can operate autonomously while still contributing to a cohesive, unified analytical ecosystem.

The Kimball architecture is built for iterative development, guided by the evolving analytical needs of the data warehouse. The process starts by identifying key facts and dimensions, followed by gathering and consolidating the necessary data in the staging area to construct the required dimensional model. As new facts and dimensions are introduced, the architecture encourages a continuous evaluation of whether these elements can be conformed to existing facts or dimensions already utilized by other data marts.

This approach promotes the reuse of existing data in the staging area, potentially extending consolidation logic to ensure alignment across domains. Only when new facts or dimensions cannot be conformed to those already present does the architecture introduce new data into the staging area for processing.

Kimball's architecture offers a clear advantage by enabling the development of interoperable data marts. In architectures where data marts are built in isolation, with each team independently sourcing and modeling data directly from operational systems, ensuring cross-domain data integrity and interoperability becomes a significant challenge. By introducing a staging area, Kimball's approach centralizes data collection and integration, creating a shared foundation of consolidated data that serves as the backbone for building interoperable data marts.

The staging area focuses purely on transformations needed to support specific data marts, allowing for fast and efficient data preparation. However, outside of conformed facts and dimensions, the reuse of consolidated data in the staging area is limited. Each new data mart typically requires starting from raw source data and re-consolidating it with new business logic tailored to the specific analytical needs.

To overcome the limitations of Kimball's two-layer architecture, Bill Inmon proposed a three-tier architecture in his book, *Building the Data Warehouse*. This approach introduces an additional layer between the staging and presentation layers, known as the **enterprise data warehouse** (**EDW**). In this model, the staging layer focuses solely on collecting raw data from various sources. The EDW then consolidates and integrates this data into a highly normalized, comprehensive model (*Figure 10.4*).

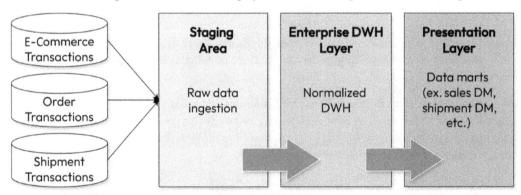

Figure 10.4 – Inmon data warehouse architecture

The EDW acts as a robust, centralized repository designed to integrate data from all relevant applications, serving as a foundation for broader analytical needs. Unlike Kimball's staging area, which is optimized for specific data marts, the EDW is built to support a wide range of analytical requirements, both current and future. This architecture facilitates the creation of new data marts by leveraging the integration work already accomplished at the enterprise level. However, developing and maintaining this intermediate layer is more costly and complex.

Having covered the basics of dimensional modeling and its application in traditional centralized architectures, let's now delve into how it can be leveraged within a decentralized architecture centered around data products.

Distributed dimensional modeling

The presentation layer, structured as a collection of related yet independent data marts in both Kimball's and Inmon's frameworks, naturally fits within a decentralized architecture centered on data products. This setup is easily adaptable to a distributed environment where each data mart can be aligned with a specific business domain, effectively representing a bounded context.

In this approach, each dimensional model and its associated data within a data mart can be managed as a data product by a dedicated team focused on that particular domain. Meanwhile, teams responsible for various source systems, or a central sourcing team within IT, can concentrate on developing source-aligned data products. These source-aligned products then feed into the consumer-aligned data products managed by the domain-specific teams (*Figure 10.5*).

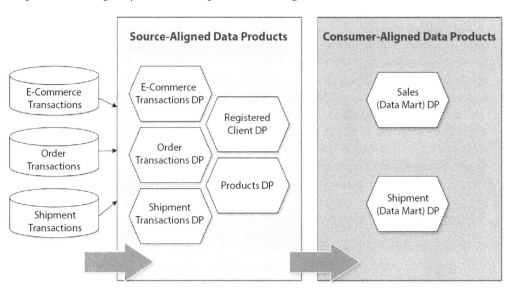

Figure 10.5 – Distributed dimensional modeling

In this data-product-based architecture, where there is no centralized intermediate layer to handle cross-source data integration and consolidation, this responsibility shifts to consumer-aligned products. This distributed approach introduces challenges in terms of data ownership, especially when dealing with conformed facts and dimensions that are critical for maintaining data integrity and interoperability across various data products:

- **Ownership of conformed dimensions**: When a conformed dimension is replicated across multiple data products, it results in creating multiple versions of the same data in different places, each with its own ownership. Ideally, each conformed dimension should be managed as a separate data product. However, this raises issues of correct ownership attribution. For example, consider the product dimension used by both sales and shipments. Not all products are shipped (e.g., products only available in-store), and not all shipped products are sold (e.g., promotional products). Additionally, the attributes relevant to sales (e.g., price) might differ from those relevant to shipments (e.g., weight). Deciding whether to assign ownership of the product dimension to the sales domain or the logistics domain is not straightforward but rather arbitrary. In either case, part of the data managed by the development team will fall outside its domain of expertise.

- **Ownership of conformed facts**: In a data-product-centric architecture, conformed facts across different data marts typically represent separate sets of data. For instance, conformed facts for in-store sales and online sales usually do not overlap, as a sale is either made in a physical store or through an online platform. Without data overlap, it's feasible for the teams managing in-store sales and those handling online sales to collaborate on defining metrics of mutual interest (e.g., units sold). Instead, metrics that are specific only to online or in-store sales can be calculated independently by the respective teams (e.g., online discount). However, there are often metrics that could potentially be calculated from both sets of facts but are meaningful and can be computed more easily by just one of the domains. For instance, suppose the online sales domain is interested in calculating the total weight of purchased goods. This metric is theoretically applicable to all purchases, not just online ones, but is of no interest to the in-store sales domain. In this case, either you compel the in-store sales domain to calculate the total weight, which places a responsibility outside their natural area of expertise and interest, or you calculate it only within the online sales domain and potentially shift the responsibility of calculating the total weight of in-store purchases to the final data consumer. Both approaches to ownership present challenges and may not be ideal.

Let's explore how the Data Vault approach can help address these ownership issues.

Data Vault modeling

The Data Vault architecture, as outlined by Daniel Linstedt and Michael Olschimke in *Building a Scalable Data Warehouse with Data Vault 2.0*, represents an evolution of Inmon's three-tier architecture, designed to tackle its key scalability and extensibility issues. Central to this architecture is the **Data Vault modeling** technique, which organizes data within the enterprise data warehouse layer. This approach revolves around three core entities:

- A **Hub** represents a business entity of interest for data collection. It contains only the natural key that uniquely identifies the entity, along with metadata, such as the source system from which the identifier was loaded and the timestamp of this load. Each object contained in a Hub has a surrogate key generated by hashing one of its natural keys. Typically, for each dimension present in the dimensional model within the presentation layer, there is a corresponding Hub in the Data Vault model within the EDW layer (e.g., customer, product, store, etc.).

- A **Link** represents a relationship between business entities. It comprises only the keys of the Hubs associated with the related entities, along with metadata, such as the source system from which the relationship was loaded and the timestamp of this load. Each relationship contained in a Link has a surrogate key generated as a hash of the natural keys of the related objects. Typically, a Link encapsulates a transaction that connects related entities within a business process. For instance, a Link might represent a sales transaction, linking the identifiers for the customer, the purchased product, and the store where the transaction occurred. In the Data Vault model, each fact in the dimensional model within the presentation layer generally corresponds to a Link in the Data Vault model located in the EDW layer.

- A **Satellite** captures the descriptive details of an entity within a Hub or a relationship within a Link. It includes the attributes that provide context to the entity or relationship, along with metadata, such as the source system from which these attributes were loaded and the timestamp of the load. Each set of attributes in a Satellite references the corresponding object or relationship in a Hub or Link by using the hash key of that Hub or Link. Whenever an attribute value changes, the new value is appended to the Satellite rather than overwriting the existing one. This ensures that both current and historical data states are preserved and accessible, providing a comprehensive view of how the data has evolved. Hubs and Links can be associated with multiple Satellites. Generally, attributes related to the same Hub or Link are grouped into Satellites based on their source systems to facilitate data ingestion.

In summary, a Data Vault model consolidates data from diverse sources into a unified, normalized model where primary keys, foreign keys, and descriptive attributes are organized into separate but interconnected entities. Hubs and Links provide the foundational structure, encapsulating primary and foreign keys from the source systems, while Satellites store all descriptive attributes (*Figure 10.6*).

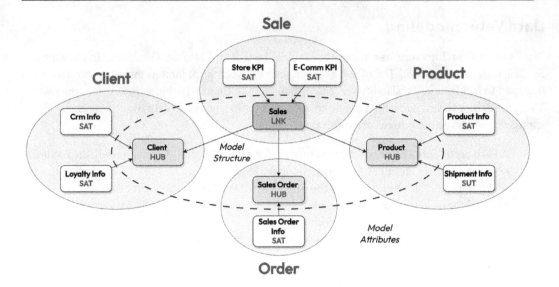

Figure 10.6 – Data Vault model

This modeling technique facilitates data integration while minimizing the impact on the model from the addition of descriptive attributes or new types of relationships between business entities introduced by new or modified source systems.

In the Data Vault architecture, the modeling approach outlined here is primarily used to construct the Raw Data Vault. This **Raw Data Vault** is focused on structuring data from source systems into Hubs, Links, and Satellites, without applying complex business transformations. It maintains the original meaning of the data, reflecting it precisely within an integrated model.

From this foundational Raw Data Vault, dimensional models can be developed to build various data marts. This process involves joining relevant Hubs, Links, and Satellites to produce the desired facts and dimensions, incorporating business logic that influences the data's quality and meaning. To streamline the process and avoid replicating business logic across multiple data marts, a **Business Data Vault** can be constructed on top of the Raw Data Vault. This Business Data Vault employs the Data Vault modeling approach but adds business logic transformations that can modify both the structure and the meaning of the data (*Figure 10.7*).

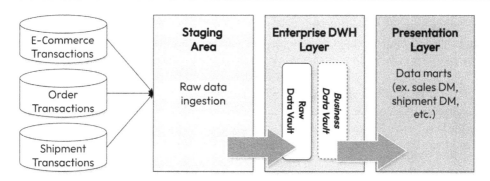

Figure 10.7 – Data Vault architecture

The Data Vault architecture encompasses a range of additional entities and optional structures beyond the ones we've discussed, such as point-in-time tables and bridge tables, as well as operational and metric data vaults. For a comprehensive understanding of Data Vault, refer to the resources in the *Further reading* section at the end of this chapter.

Now, let's explore how Data Vault modeling can effectively address some of the ownership attribution challenges associated with dimensional modeling.

Distributed Data Vault modeling

The Data Vault modeling technique offers a robust framework for structuring data exposed by source-aligned products. In this approach, each product is responsible for managing its own specific Satellite, while Hubs and Links are centralized and managed at the platform level.

Managing Hubs and Links as a service

When a source-aligned data product updates its data, the platform synchronizes by updating the associated Hub or Link associated with the Satellite managed by that data product. This synchronization is designed to be both straightforward and automated. The data product sends the platform the natural keys of the objects or relationships added during the latest load. The platform then integrates these keys—if they aren't already present—into the corresponding Hub or Link. Subsequently, it returns the hash keys that identify the updated objects or relationships to the data product.

The data product then updates its internal objects or relationships with these hash keys, replacing the natural keys for each item with the hash keys, as specified by Data Vault modeling principles. While this process is efficient, in distributed environments, it might be worth considering retaining the natural keys alongside the hash keys. This practice ensures that the data exposed by the data product remains comprehensible and usable, even if the platform managing Hubs and Links is unavailable.

When a Satellite is associated with a Link, updating the Link with new data from the Satellite might trigger not only updates to the Link but also cascading updates to the associated Hubs. This situation arises when attributes are added to a relationship involving objects that were not previously registered in the corresponding Hubs. In these scenarios, the data product faces a choice: it can either register the objects in the relationship with the platform first and then register the relationship itself, or it can directly register the relationship and leave the task of documenting any new objects to the platform. Typically, the first approach is favored when the objects involved in the relationship come from different systems and there is a need to maintain traceability of the data sources at the Hub level.

Decentralizing ownership across Satellites

Now, imagine managing data for sales and shipments using distributed data modeling as described here. The data marts in the consumption layer will naturally share several dimensions, such as product or customer, that need to be conformed. The approach to handling these dimensions through dedicated data products ensures that multiple copies of the same data, each with distinct ownership, are avoided. When leveraging Data Vault modeling in source-aligned products, managing ownership attribution for these conformed dimensions also becomes more streamlined.

Take, for instance, the product dimension. Information about products in the catalog may come from various source systems, each under different domain responsibilities. The sales domain might be responsible for maintaining descriptive attributes related to the product's main characteristics—color, size, brand, and so on. Conversely, the logistics domain might focus on attributes essential for shipping—weight, packaging type, and so on. In Data Vault modeling, these two types of information can be managed independently through separate satellites linked to the same product Hub and managed by independent data products. The sales domain would oversee the data product managing the Satellite with general product information, while the logistics domain would manage the Satellite with shipping-related information. The platform supports this by providing a centralized Hub containing the natural keys for products (e.g., **stock-keeping units (SKUs)**) to which both Satellite data products are correlated through the hash key (*Figure 10.8*).

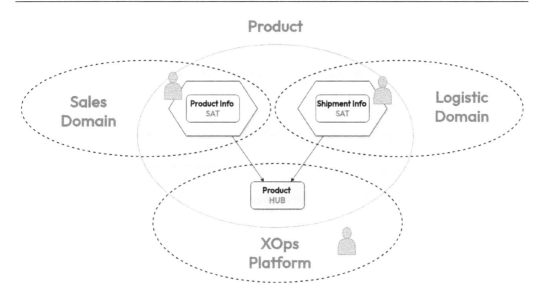

Figure 10.8 – Distributed Data Vault modeling

The same applies to conformed facts. Take, for instance, the sales fact, which encompasses both in-store transactions and online sales. Here, we'd establish two distinct data products: one managed by the in-store sales domain and the other by the online sales domain. Each data product would oversee two satellites correlated to the sales Link—one Satellite containing the conformed measures that it will populate only for its own sale transactions, and one containing metrics relevant only to the specific domain that it could potentially populate for all sales transactions. As with Hubs, the Link is managed by the platform and referenced by the Satellites managed by the two data products.

In a robust data architecture, it's crucial to maintain satellites that preserve raw data sourced directly from the original systems. These satellites, managed by source-aligned data products, collectively build the Raw Data Vault.

When complex business rules are needed to alter the data's meaning, it is preferable to apply these transformations within dedicated satellites. This approach also holds for KPIs calculated from source system metrics. Such satellites built applying business logic to transform and enrich the row data can be managed by specialized consumer-aligned data products that are constructed downstream from the source-aligned products. Together, these enriched Satellites form the **Business Data Vault**.

Moving toward the consumption layer

With the source-aligned products that expose Satellites within the Raw Data Vault and the downstream data products that expose Satellites within the Business Data Vault in place, you can leverage the integrated model to generate facts and dimensions for the desired data marts. This is achieved through simple joins between Hubs, Links, and Satellites.

These data marts and their associated data function as **aggregation data products**, seamlessly integrating data from various upstream products to offer a unified, consumable view of a business process or transaction. The lack of transformation logic enables these data products to be virtualized, with ownership assigned to the federated modeling team—described in *Chapter 11, Building an AI-Ready Information Architecture*—rather than to domain-specific teams.

In these consumer-aligned virtualized data products, data has not materialized and is so not duplicated, except for performance optimization in queries. The responsibility for each piece of data they use remains with the upstream data product team that provides it. The federated modeling team is only responsible for the logic used to join the data to produce the exposed data mart. In the event of data issues, it is easy to identify who is accountable for resolving them.

Unified star schema modeling

The **unified star schema** (**USS**) offers a modeling approach for data within the presentation layer, designed to streamline the structure and improve the querying experience by eliminating errors caused by join traps.

Join traps

Each data mart within the presentation layer is composed of one or more dimensional models. Each dimensional model is structured using either a star schema or a snowflake schema and is managed by a consumer-aligned data product, typically virtualized and overseen by the federated modeling team.

From each Link within the EDW layer, multiple-dimensional models can be derived in the presentation layer. However, given that the data exposed through these models are treated as products, it's typically recommended to have no more than one model per Link. Consumers can then connect, filter, and aggregate the models as needed to obtain the desired data form.

However, creating a single-star schema for each Link, encompassing all relevant data at the finest level of granularity, is not always feasible due to the inherent risks of join traps. A **join trap** occurs when a join operation inadvertently duplicates rows within a table, leading to inaccuracies in aggregate measure calculations.

The star schema is inherently designed to address this issue. By structuring data so that all measures reside within a fact table, which maintains a many-to-one relationship with its dimensions, the risk of duplication is minimized. This design ensures that when joining the fact table with its dimensions, each row in the fact table remains unique, preserving the integrity of the measures it holds. Consequently, calculations of aggregate measures within a single-star schema—often referred to as **drill-down**—are safeguarded from errors arising from join traps.

The challenge arises during a **drill-across** operation, where the risk of join traps re-emerges. In this scenario, joining two distinct fact tables can lead to multiple duplications of rows across both tables and, consequently, their measures. This can result in aggregation errors. For instance, consider the scenario involving the sales fact and the shipment fact (*Figure 10.9*).

SalesID	Client	SalesDate	ProductID	SalesQuantity	SalesAmount
1	A	09-Jan	PR01	1	100
2	A	09-Jan	PR02	1	70
3	B	09-Jan	PR02	2	140
4	B	09-Jan	PR03	1	300
5	B	09-Jan	PR01	40	4,000
				45	4,610

ShipmentID	SalesID	ShipmentDate	ProductID	ShipmentQuantity	ShipmentAmount
1	1	09-Jan	PR01	1	100
2	2	09-Jan	PR02	1	70
3	3	09-Jan	PR02	2	140
4	4	09-Jan	PR03	1	300
5	5	09-Jan	PR01	10	1,000
6	5	09-Jan	PR01	30	3,000
				45	4,610

Figure 10.9 – Sales and shipments facts

Connecting sales and shipment data for comparative analysis or to derive metrics, such as the daily delta between units purchased and shipped for each product or the monthly ratio of sales amounts to shipping costs, often makes sense. The relationship between sales and shipments is typically direct: each shipment is potentially tied to a specific sale via a foreign key linking shipments to sales.

Yet, it's crucial to recognize that not all purchased units are necessarily shipped in a single batch. Consequently, the same sales record might appear multiple times in the join result, once for each corresponding shipment. This duplication introduces the risk of incorrect aggregations: while the units shipped will aggregate accurately, the units sold will be miscalculated due to repeated sales rows. This scenario is an example of a **fan trap**—a join trap that occurs when merging entities connected by a one-to-many relationship, where both entities contain measures (*Figure 10.10*).

SalesID	SalesDate	Client	SalesQuantity	SalesAmount	ProductID	ShipmentID	ShipmentDate	ShipmentQuantity	ShipmentAmount
1	09-Jan	A	1	100	PR01	1	09-Jan	1	100
2	09-Jan	A	1	70	PR02	2	09-Jan	1	70
3	09-Jan	B	2	140	PR02	3	09-Jan	2	140
4	09-Jan	B	1	300	PR03	4	09-Jan	1	300
5	09-Jan	B	40	4,000	PR01	5	09-Jan	10	1,000
5	09-Jan	B	40	4,000	PR01	6	09-Jan	30	3,000
			85	8,610				45	4,610

Figure 10.10 – Fan trap

In rarer scenarios, it might be worthwhile to join fact tables, which aren't directly linked, through a shared dimension. Consider a case where source systems fail to capture the connection between a shipment and the corresponding sale, leading to no direct relationship between the two facts. Despite this, it's still possible to establish a correlation, such as calculating the daily delta between units shipped and units sold, by leveraging the common product dimension shared by both facts.

However, this approach can introduce row duplications on both sides of the join. Each sale of a product will be repeated as many times as there were shipments of that product, and each shipment of a product will be repeated as many times as there were sales of that product. This type of join trap is called a **chasm trap** (*Figure 10.11*).

SalesID	Customer	SalesDate	SalesQuantity	SalesAmount	ProductID	ShipmentID	ShipmentDate	ShipmentQuantity	ShipmentAmount
1	A	09-Jan	1	100	PR01	1	09-Jan	1	100
1	A	09-Jan	1	100	PR01	5	09-Jan	10	1,000
1	A	09-Jan	1	100	PR01	6	09-Jan	30	3,000
5	B	09-Jan	40	4,000	PR01	1	09-Jan	1	100
5	B	09-Jan	40	4,000	PR01	5	09-Jan	10	1,000
5	B	09-Jan	40	4,000	PR01	6	09-Jan	30	3,000
2	A	09-Jan	1	70	PR02	2	09-Jan	1	70
2	A	09-Jan	1	70	PR02	3	09-Jan	2	140
3	B	09-Jan	2	140	PR02	2	09-Jan	1	70
3	B	09-Jan	2	140	PR02	3	09-Jan	2	140
4	B	09-Jan	1	300	PR03	4	09-Jan	1	300
			130	13,020				89	8,920

Figure 10.11 – Chasm trap

Most **business intelligence** (**BI**) tools elegantly sidestep these join traps using **associative logic**. When mapping the model within the tool and establishing how tables are connected, the tool resolves these connections via associations rather than traditional joins. This means that the tool loads the individual tables into memory and then resolves the associations only when necessary, calculating measures accurately on the individual tables before performing the joins.

However, direct data access, such as through SQL queries, does not benefit from this associative functionality. To mitigate the risk of join traps in such scenarios, it is common practice to create multiple-star schemas tailored to different types of cross-fact analysis. While this approach addresses potential pitfalls, it inevitably leads to an increase in the number of star schemas, thereby escalating the complexity and costs associated with managing the presentation layer.

Puppini Bridge

The USS is a data modeling approach designed to tackle these challenges within the presentation layer. As outlined by Francesco Puppini and Bill Inmon in their book, the USS revolves around a pivotal concept: consolidating multiple consumption needs into a single fact table rather than creating a distinct fact table for each requirement. This central fact table, referred to as the **Puppini Bridge**, serves as the broker through which all other relevant tables are interconnected.

The Puppini Bridge includes a column for each key from the tables you want to correlate. If any of these tables have composite keys, a single key must be created—often by hashing the composite key—for use in the Puppini Bridge. Along with these keys, the Puppini Bridge also contains columns for each measure from the different correlated tables and an additional column called `Stage`.

Once the schema is set up, the Puppini Bridge is populated by doing the union of all rows coming from all the tables being correlated. For each row, it stores the primary key, foreign keys, measures, and the name of the table in the `Stage` column. The following figure shows an example of a Puppini Bridge constructed from the shipment, sales, and product tables (*Figure 10.12*).

Stage	ProductID	SalesID	ShipmentID	SalesQuantity	SalesAmount	ShipmentQuantity	ShipmentAmount
Product	PR01						
Product	PR02						
Product	PR03						
Sales	PR01	1		1	100		
Sales	PR02	2		1	70		
Sales	PR02	3		2	140		
Sales	PR03	4		1	300		
Sales	PR01	5		40	4,000		
Shipment	PR01		1			1	100
Shipment	PR02		2			1	70
Shipment	PR02		3			2	140
Shipment	PR03		4			1	300
Shipment	PR01		5			10	1,000
Shipment	PR01		6			30	3,000
				45	4,610	45	4,610

Figure 10.12 – Puppini Bridge

At this point, you can simply perform a `LEFT JOIN` operation between the tables of interest and the Puppini Bridge to analyze the contained data, effectively sidestepping join traps. Fan traps and chasm traps are avoided because the Puppini Bridge connects tables through union, not join, thus preserving their original granularity. The join operations to retrieve the necessary attributes for slicing and dicing occur only during the calculation phase.

Since the Puppini table holds all the measures of interest and maintains a many-to-one relationship with other tables joined during calculations, it ensures that each measure is counted only once, avoiding any risky duplication.

Of course, the Puppini Bridge can have high cardinality. However, it is a highly sparse table that contains only keys and measures. All informational attributes remain in their respective tables, ready to be used during calculations as needed. As a result, the storage required relative to cardinality is kept manageable. Moreover, since it can be easily partitioned and indexed, both storage and read performance can be further optimized.

Now, let's delve into how the USS technique can be leveraged within a distributed architecture that centers around data products.

Distributed unified star schema modeling

Starting with data in the EDW layer that is modeled using the Data Vault approach, constructing a Puppini Bridge is relatively straightforward. In Data Vault modeling, primary and foreign keys, stored in Links and Hubs, are already distinct from the descriptive attributes held in Satellites. Each element in a Hub or Link also has a non-composite surrogate key, generated by hashing one of its natural keys. This setup allows for the easy creation of a Puppini Bridge by merging the hash keys from the various Links and Hubs. Additionally, for each Link, the Puppini Bridge needs to include the relevant measures from the associated Satellites.

While it's technically possible to create a single Puppini Bridge that connects all Hubs and Links from a Data Vault-modeled **data warehouse** (**DWH**) layer, doing so is often neither necessary nor efficient. Instead, Hubs and Links can be grouped by major analytical areas, allowing for the creation of dedicated Puppini Bridges tailored to each area.

Since a Puppini Bridge can be automatically built from Links, Hubs, and Satellites, its creation can be offered as a self-service feature by the platform and executed on-demand based on consumption needs. For instance, when a consumer selects data products within the data product marketplace, the platform can automatically provide access not only to the individual data products but also to a dedicated Puppini Bridge for safely joining data across those products.

Next, let's explore how to manage the life cycle of physical models in a distributed architecture centered around data products.

Managing the physical model life cycle

Up to this point, we've delved into the management of physical data modeling within data-product-centric architectures. Specifically, we've advocated for employing Data Vault techniques to organize data exposed by source-aligned data products, and for leveraging star schema or USS approaches to model data presented by consumer-aligned products (*Figure 10.13*).

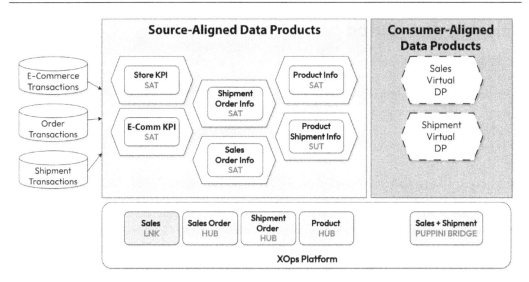

Figure 10.13 – Distributed modeling architecture

We have also explored how to build a unified model in a distributed manner for each described technique, paying close attention to aspects of ownership, reduction of interdependencies, and interoperability, among different parts of the model. Each of these parts is managed by a different product team and is an integral part of the product itself, thereby following its life cycle.

A distributed data model emerges through the composition of individual data product models, each developed autonomously. The data model is not a separate entity but an intrinsic part of the data product itself, evolving alongside it throughout its life cycle. Every data product is responsible for offering a transparent and well-defined representation of its internal data model, ensuring consumers can easily understand how the data is structured and exposed. This model description is embedded within the data product's descriptor and must remain readily accessible through its discoverability ports.

The construction policies and representation standards for data models are set by the federated governance team. Product teams, tasked with building data models, must align with these globally defined rules. The XOps platform plays a critical role here, ensuring that these rules are computationally validated during the deployment process, with the authority to block non-compliant data products. Furthermore, the XOps platform can leverage the information present in the physical data models of various data products to simplify search tasks or automate procedures that facilitate data consumption (e.g., generating a Puppini Bridge on demand).

The data model exposed by a data product serves as one of the most vital interfaces that binds the product to its consumers. Introducing changes to this model in new product versions can disrupt downstream consumers, making it crucial to manage these changes with caution. Any alterations to the model should be handled in alignment with the deprecation policies that govern the entire life cycle of the data product, ensuring that consumer impact is minimized while maintaining a consistent and reliable interface for data access.

Exploring distributed conceptual modeling

In traditional approaches, the conceptual model is often loosely defined during the analysis phase to guide the construction of the physical model. However, in a data-product-centric architecture, the conceptual model plays a critical role well beyond the initial release of data products. It serves as a key enabler for seamless data consumption and ensures interoperability across products. In this section, we'll explore how to intentionally design a conceptual model and manage its ongoing evolution in a distributed decentralized architecture.

Ubiquitous language

The primary purpose of conceptual modeling is to surface the core concepts that shape the problem and map out the relationships between them. Everyone involved in solving the problem should participate in the modeling activity, which typically aligns with the early stages of analysis to identify the necessary data products to support a given business case. It's especially critical that business stakeholders with deep knowledge of the processes related to the target problem participate in the conceptual modeling, as they hold most of the knowledge needed to construct this kind of model.

A modeling session kicks off with a deep dive into the business processes tied to the business case in question. This process analysis can unfold in several ways, such as through techniques such as event storming (see *Chapter 4, Identifying Data Products and Prioritizing Development*). As you delve into the process, it's essential to capture the most frequently used terms, especially those highlighted by subject matter experts. These terms embody the language specific to the bounded context you're working within.

For each significant term related to the process, a clear and shared definition should be formalized. If multiple terms are used to describe the same concept, they should be tracked and a preferred term should be selected. This preferred term will then serve as the foundation for the **ubiquitous language** used across all artifacts delivered to achieve the desired solution.

The set of key terms identified, along with their definitions and potential synonyms, is classified within **knowledge organization systems (KOS)** as a **controlled vocabulary**.

Establishing a shared controlled vocabulary is crucial for eliminating misunderstandings between those engaged in both the analysis and the implementation of the desired solution. For example, when the term "customer" is mentioned, stakeholders can refer to a common definition, ensuring everyone is aligned on the meaning and context of the information being exchanged.

The controlled vocabulary plays a crucial role even after development, serving to document the physical model and enhance data comprehension for potential consumers. This is achieved by tagging the schema—fields, tables, and other data structures—with terms from the controlled vocabulary. When consumers interact with the physical model via the product's discoverability ports, they gain not only structural insights but also contextual understanding of the data. The tags function as links to the shared controlled vocabulary, providing access to term definitions and thus clarifying the meaning of the tagged data.

Controlled vocabularies are designed to be shared across multiple data products, especially within the same bounded context or subdomain. This approach avoids redundant definitions of terms, as the meaning of these terms remains consistent across products.

While the controlled vocabulary evolves alongside the development of the data products that utilize it, it operates independently of them and does not share its life cycle. The controlled vocabulary is neither owned nor exposed by any specific data product. Unlike the physical model, which is directly exposed through the data product's discovery ports, the controlled vocabulary must be managed externally—either by the XOps platform or a connected data governance tool. The only responsibility of the data products is to tag the exposed physical model appropriately and ensure these tags are available for consumers to use.

In data governance platforms, the management of a controlled vocabulary is often facilitated through a component known as the **business glossary**. This Business Glossary acts as a centralized, user-friendly repository where vocabulary terms are cataloged, linked, and maintained over time. It serves as a critical tool for both understanding and navigating the data landscape.

Through the business glossary, users can not only quickly access the definition of a term used to tag data for better understanding but also search for terms within the vocabulary and trace them back to associated data. In fact, the business glossary indexes terms and their definitions, enabling text-based searches within the controlled vocabulary. Additionally, it keeps track of all physical models tagged with specific terms. This allows you, for instance, to find all terms related to the search string `Customer`. Then, by selecting the most relevant result, such as `Active Customer`, users can view its definition and see all physical models tagged with this term, thus identifying the data products exposing these models and the associated data.

It is the responsibility of the XOps platform to read the tags used by the data product each time a new version is released and update the mappings between terms and data products within the business glossary.

As we've discussed, a controlled vocabulary is an invaluable tool for conceptual modeling, essential during both the development and operational phases of a data product. Formalizing the vocabulary during the analysis phase—when all subject matter experts are present and actively involved—is more effective and cost-efficient than doing so later through post-release stewardship activities. However, while formalizing and cataloging the controlled vocabulary within the business glossary is crucial, it does not complete the conceptual modeling process. The controlled vocabulary is just the starting point, a foundational tool for creating more expressive models with capabilities beyond simple textual search. In the next section, we will delve into these advanced models and how to define them.

From string to things

The controlled vocabulary provides a flat list of terms with their definitions and key synonyms but lacks structure. It does not distinguish concepts from their properties or define the relationships between concepts. This more structured approach, which organizes concepts, properties, and their interconnections, is called an **ontology**. Ontologies allow for a wide range of relationships to link concepts, making them the most versatile and comprehensive form of explicit representation for shared knowledge.

In knowledge management systems, simpler forms of representation with respect to ontologies are often used, typically built from a more limited set of possible relationships between objects. The most common of these are taxonomies and thesauri. **Taxonomies** establish a hierarchical relationship between concepts or terms from the controlled vocabulary using relationships, such as "is broader than" and "is narrower than." This creates a hierarchy where each concept or term is linked to a parent term in a structured, hierarchical manner. On the other hand, a **thesaurus** supports both hierarchical and associative relationships. It allows terms and concepts to be linked across different taxonomies, enabling a more flexible and interconnected representation of knowledge.

A thesaurus can be viewed as a specific type of ontology that employs only hierarchical and associative relationships. Similarly, a taxonomy can be seen as a particular kind of thesaurus where only hierarchical relationships are used. Finally, a controlled vocabulary can be considered a specific type of taxonomy in which terms are not interconnected by any relationships. Generally, as illustrated in the following figure, ontologies are composed of thesauri, which link taxonomies, which in turn link the terms found in a controlled vocabulary (*Figure 10.14*).

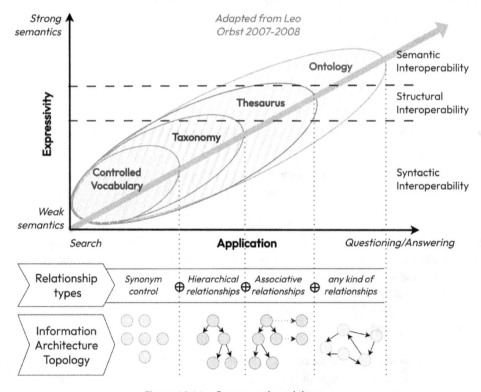

Figure 10.14 – Conceptual model types

The more complex the structure of a conceptual model, the more expressive it becomes. In these models, a term's meaning isn't defined solely by the word or its textual definition in the vocabulary, but by its position within the structure and its relationships to other terms. In a well-designed conceptual model, the textual definition of a concept should be redundant because its meaning is embedded in the structure itself. Take the term "customer," for instance. In a controlled vocabulary, "*customer*" might be defined as *A customer is an individual or entity that purchases products or services from a business in exchange for monetary value*. The illustration (*Figure 10.15*) shows a hypothetical ontology that represents the concept of *Customer*, transforming its textual definition ("Strings") into a structure of related concepts ("Things").

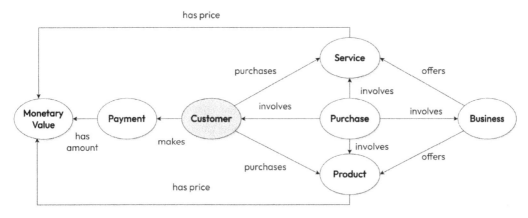

Figure 10.15 – Ontology

This ontology conveys the same informational content as the previous definition but is less prone to ambiguity and more easily processed by machines. While the textual definition is useful for performing searches across defined terms, the ontology allows for reasoning through inferences based on its graph structure. For instance, if it's known that a customer purchased a product, and every product has a price, the payment amount can be inferred even if it's not explicitly recorded in the system. Similarly, if the recorded payment differs from the product price, it could indicate a data quality issue.

Managing the conceptual model life cycle

An ontology is a purposeful model crafted to capture domain knowledge with the intent of addressing specific problems. The journey of modeling an ontology begins with a clear identification of the target problem, the business outcomes we seek to achieve, and the business questions the model must help answer. Conceptual modeling should never be approached as an abstract, exhaustive mapping of all organizational knowledge. This would amount to little more than an intellectual exercise, disconnected from value creation. Instead, the process must remain tightly aligned to real business cases, ensuring the model directly serves actionable needs.

With the motivations and scope of the modeling clearly defined, the next step is to bring together the subject matter experts who hold deep knowledge of the problem domain and the associated business processes. This initiates the first phase of conceptual modeling. The focus here is to uncover the core concepts that form the foundation of the model. The process often begins by analyzing business processes related to the target problem. As key terms emerge, it's crucial to assess whether they represent concepts, properties of those concepts, or relationships between them, and then document them in the shared vocabulary. As more concepts are identified, they can also be visualized within a graph-based model, representing their key properties and the connections that define their relationships.

Once the initial process analysis is completed and an informal conceptual model is drafted, the next step involves refining the understanding of key concepts. This phase focuses on better identifying the attributes that describe these concepts and the relationships that connect them to others.

The construction of the ontology unfolds in parallel with the definition of the controlled vocabulary. Both processes should avoid striving for perfection—such a model does not exist. Instead, the goal is to create a model that is lean and functional, tailored to answering the specific business questions that frame the problem. When the model achieves this goal, the collaboration with subject matter experts can be closed, and the consolidation phase can commence.

In the consolidation phase, the model undergoes a transformation to more effectively represent the collected information, making it clearer and more accessible. This phase also involves formalizing the model into a machine-readable format, typically using the **Resource Description Framework (RDF)** standard. RDF organizes information into triples, consisting of a subject, predicate, and object. The subject represents the concept being described (e.g., Customer, Product), the predicate denotes a property or relationship associated with the subject (e.g., has name, has price), and the object is either a value or another concept related to the subject through the predicate (e.g., Andrea, Monetary Value). The following examples illustrate a couple of RDF triples derived from the ontology presented in the previous figure:

```
(Customer, purchases, Product)
(Customer, purchases, Service)
```

After logically structuring the information within the ontology using RDF triples, these triples can be serialized into various formats for consumption. Among the available formats, **Turtle** stands out for its compactness and readability. It offers a human-friendly way to represent RDF triples. The following is an example demonstrating how the previous RDF triples can be serialized using the Turtle format:

```
:purchases a rdf:Property ;
    rdfs:domain :Customer ;
    rdfs:range [ a owl:Class ;
        owl:unionOf ( :Product :Service ) ]
```

In this example, the Turtle syntax is used to specify that the purchases predicate accepts Customer as its subject (domain) and either Product or Service as its object (range).

Ontologies that are defined, formalized, and serialized in this way can be stored and accessed through specialized databases that natively handle triples (triplestores), graph databases, or governance tools capable of managing ontologies.

The controlled vocabulary remains a crucial tool as it annotates the ontology in a format that is more immediately understandable for business users. However, data products no longer use the vocabulary terms for tagging the physical model; instead, they create direct links to the concepts within the ontology. The tool managing the ontology must store these pointers to enable quick navigation from a concept to the data products that expose instances of that concept. We'll delve more into the intricacies of linking physical data with conceptual models in the upcoming chapter.

As with the controlled vocabulary, it is the responsibility of the federated governance team to define the rules and standards for constructing an ontology. The ontology is developed and evolved alongside the development of data products but is not part of them and does not share their life cycle. The data product is solely responsible for linking the exposed data to the appropriate concepts within the ontology and for making these links available through its discoverability ports. Instead, the XOps platform must process these links whenever a new data product is published, cataloging them within the system used to store ontologies.

Creating and evolving a shared formal ontology involves costs; at first glance, it might appear that sticking to informal conceptual modeling, which is sufficient for physical model creation, is more practical. However, as the next chapter will show, this investment is justified. The ontology ensures long-term interoperability between data products and enhances the effectiveness of modern AI techniques.

Summary

In this chapter, we explored how to model data formally and intentionally within a data-product-centered architecture. We focused on analyzing how key techniques for physical and conceptual data modeling can be applied within a modular and potentially distributed data management solution.

Initially, we examined how, for data products, data modeling transcends mere development to become an intrinsic part of the product itself. The physical model is the gateway through which consumers interact with the product, accessing and utilizing the exposed data. Conversely, the conceptual model provides the framework for understanding the meaning of that data, enabling proper usage and integration.

We then reviewed the main techniques for physical data modeling, particularly for analytical purposes. For each technique, we assessed the advantages and disadvantages, exploring their applicability within distributed data management architectures. We proposed a two-tier architecture, suggesting the use of Data Vault modeling for source-aligned data products and dimensional modeling in the form of a star schema or USS for consumer-aligned data products.

Finally, we addressed conceptual modeling, analyzing the primary techniques from controlled vocabularies to ontologies. For each technique, we discussed its strengths and weaknesses and how to define, manage, and evolve them over time.

In the next chapter, we'll delve into integrating conceptual and physical modeling to construct a comprehensive information architecture grounded in the knowledge graph concept. We'll explore how this approach supports both the composability of data products and the application of modern AI techniques.

Further reading

For more information on the topics covered in this chapter, please see the following resources:

- *My Definition of Data Modeling (for today).* J. Reis (2024). `https://joereis.substack.com/p/my-definition-of-data-modeling-for`

- *The Align, Refine and Design Approach to Data Modeling.* S. Hoberman. `https://www.youtube.com/watch?v=-5uZlyAn3Vk`

- *The Data Dichotomy: Rethinking the Way We Treat Data and Services.* B. Stopford (2016). `https://www.confluent.io/blog/data-dichotomy-rethinking-the-way-we-treat-data-and-services/`

- *Data on the Outside vs. Data on the Inside.* P. Helland (2020). `https://queue.acm.org/detail.cfm?id=3415014`

- *The Data Warehouse Toolkit: The Definitive Guide to Dimensional Modeling.* R. Kimball, M. Ross (2013). `https://a.co/d/dOnBzxJ`

- *Building the Data Warehouse.* W. H. Inmon (2005). `https://a.co/d/efuuDO6`

- *Building a Scalable Data Warehouse with Data Vault 2.0.* D. Linstedt, M. Olschimke (2015). `https://a.co/d/cNURox4`

- *The Unified Star Schema: An Agile and Resilient Approach to Data Warehouse and Analytics Design.* B. Inmon, F. Puppini (2020). `https://a.co/d/eZAbGyY`

- *Semantic Modeling for Data: Avoiding Pitfalls and Breaking Dilemmas.* P. Alexopoulos (2020). `https://a.co/d/8WQDBJa`

- *RDF 1.1 Concepts and Abstract Syntax.* `https://www.w3.org/TR/rdf11-concepts/`

- *RDF 1.2 Turtle.* `https://www.w3.org/TR/rdf12-turtle/`

11

Building an AI-Ready Information Architecture

In the previous chapter, we explored how to model data at both the physical and conceptual levels in a distributed architecture centered around data products.

In this chapter, we will learn how to build an entire information architecture that maximizes the potential value of the managed data assets, starting from the developed data products.

We will begin by examining the layers that make up the information architecture and how each layer enriches the managed data with contextual information that helps to better understand its meaning and possible uses within a specific domain.

We will then focus on the knowledge plane, which is the level within the information architecture where data is described using interconnected conceptual models shared across the organization to ensure semantic interoperability between data products.

Finally, we'll explore how federated modeling teams can define knowledge models incrementally based on business needs and connect them to physical data, forming an enterprise knowledge graph that is essential for unlocking the full potential of modern generative **artificial intelligence** (**AI**).

This chapter will cover the following main topics:

- Exploring information architecture
- Managing enterprise knowledge
- Building an enterprise knowledge graph
- Leveraging modern AI

Exploring information architecture

Managing data assets isn't enough; building a solid information architecture is key to maximizing their value and simplifying their use. But what exactly is it, and how does it help? Let's find out in this section.

Data assets

Data is one of the most valuable assets companies have in today's economy to compete and grow in the market.

The control of money and the control of things no longer dominate the economy. Today, the real control rests with the flow of information and knowledge. The means of production have shifted to those who can handle and process information. From being organized around the flow of things and the flow of money, the economy is being organized around the flow of information. Increasingly, businesses are knowledge-based and knowledge-driven. (Peter Drucker)

According to the **International Accounting Standards Board** (**IASB**), an asset is a "resource controlled by an entity as a result of past events, from which future economic benefits are expected to flow to the entity." An asset's value can come from its use (**value in use**) or its sale (**value in exchange**). Data is a resource owned by the organization that generates it through its transactional applications, and its value is primarily tied to how it's used. However, not all data produced by these applications is usable. Some data may be irrelevant to the organization's needs, while other data may lack the necessary accuracy at the source, making it unsuitable for potential uses. Therefore, not all data is an asset. Only data with potential value in use qualifies as an asset. Data without such use value is merely a liability.

To unlock the value of data assets, the first step is, therefore, to identify which data truly has potential value in use and to implement a management system that ensures the necessary levels of data accuracy for its practical usage, both during collection and storage. Once the relevance and accuracy of the collected data are secured, it becomes essential to make it easily accessible and usable. This is where information architecture comes into play. Let's take a closer look at what that entails.

Information architecture pyramid

The word *data* comes from the Latin *datum*, which means *something given*. In its raw form, data is just a collection of unprocessed facts or figures (e.g., numbers, texts, images). Without context, data can be difficult to understand and, therefore, not very useful.

The sequence of numbers displayed in the following figure is an example of raw data (*Figure 11.1*).

31	32	33	31	31	32	31

Figure 11.1 – Data

Without context, it's challenging to grasp what these numbers represent and how to use them effectively. As a result, the raw sequence of numbers holds little value on its own.

Properly processed, structured, and contextualized data becomes information. The word *information* comes from the Latin *informare*, which means *to give form to* or *to shape*. The following table (*Table 11.1*), for instance, offers structure and context to the previous sequence of numbers, enabling us to start grasping its meaning.

Weather Forecast				
Average Temperature (°C)	Year	Month	Day	City
31	2024	July	1	Milan
32	2024	July	2	Milan
33	2024	July	3	Milan
31	2024	July	4	Milan
31	2024	July	5	Milan
32	2024	July	6	Milan
31	2024	July	7	Milan

Table 11.1 – Weather forecast information

It is the data used to describe the numerical sequence that provides the essential context needed to comprehend its meaning. Data that describes other data is commonly referred to as **metadata**. In *Table 11.1*, the metadata outlines the structure of the table, clarifying both the meaning of the numerical series (average temperature in degrees Celsius) and of other related data that offer additional context (day, city). In addition to describing the structure of the data, metadata can also detail other operational characteristics of the data, such as its origin, quality, update frequency, and consumption methods. Metadata, in all its forms, transforms raw data into comprehensible and usable information.

Using data is a necessary condition for it to express its potential value, but it is not sufficient on its own. Simply utilizing the data isn't enough; it must also generate tangible value for the organization, aligned with its strategic objectives. To achieve this, it's essential to correlate the information collected with the specific domain knowledge relevant to the organization.

Knowledge is a model of the domain of interest, developed over time through past experiences and learning. With knowledge, we can not only grasp the general meaning of the collected information but also uncover its deeper significance for the organization. This understanding enables us to contextualize the information within the organization's reality and use it effectively to drive positive outcomes.

Every individual within an organization has their own mental model of the domain in which they operate. Since each person views the organization from different perspectives, these mental models are personal and distinct. Consequently, different individuals may interpret the same set of information in various ways and act, accordingly, often basing their decisions on partial or potentially inaccurate knowledge. For this reason, as discussed in the previous chapter, it is crucial to gather, articulate, formalize, and share knowledge at the organizational level to facilitate a common understanding of collected information.

The ontology shown in the following figure (*Figure 11.2*) represents the knowledge related to a portion of the domain in which a hypothetical company called *Food Corp* operates.

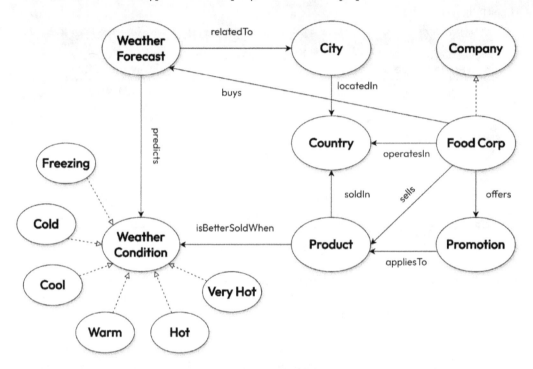

Figure 11.2 – Food Corp domain knowledge

From the model, it is clear that Food Corp is a multinational company operating in various countries in which it sells its products. Weather conditions can influence the sales of some of these products. Consequently, Food Corp purchases weather forecasts for all major cities in the countries where it operates.

The information in the previously shown table likely represents the weather forecasts acquired by Food Corp for the first week of July in Milan. The table, its structure, and the associated metadata have helped clarify the meaning of the collected data. However, it is only through the conceptual model that we can move beyond simple data comprehension to achieve a deeper, more holistic understanding. This conceptual model allows us to connect the information to Food Corp domain knowledge, see the bigger picture, and effectively analyze and use the collected data to create value for the organization.

In the absence of a shared conceptual model, each individual might interpret the information represented by the table differently. For some, this might indicate that it will be hot in Milan during the first week of July, while for others, it could imply that this July will not be particularly warm for a typical summer in Milan. Generally, the significance of this information for Food Corp may not be clear to everyone. However, with a shared model in place, it becomes evident that this information is important, as the temperatures will affect the sales of certain products.

At this stage, an intelligent agent—be it human or otherwise—can leverage the information gathered alongside the shared conceptual model to decide what to do to create value for the organization. For instance, if it's known that temperatures will be high, one valuable action could be to plan promotions on products that typically sell well in hot weather conditions. This strategy could help boost sales of these products and increase their market share.

Data, information, knowledge, and intelligence—or wisdom—constitute the four levels that shape the information architecture (see *Figure 11.3*). Each level builds on the one before it, enriching the context surrounding the collected data. The more context that is provided, the deeper and more nuanced the understanding of the data's meaning becomes, ultimately guiding its effective utilization within the organization.

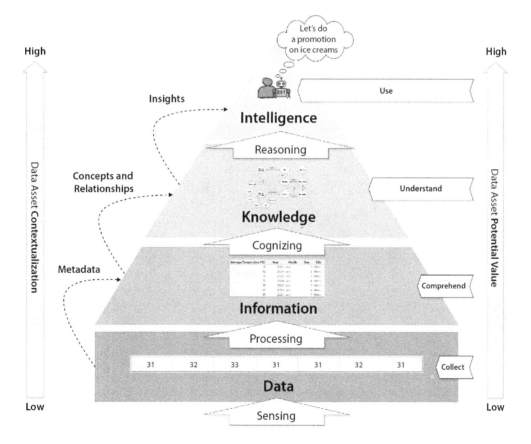

Figure 11.3 – Information architecture pyramid

To ensure that data is a real asset rather than just a liability, it is essential to manage the entire information architecture, not only the data layer. In this chapter, we will see how. Let's begin by exploring the information and knowledge planes in the following sections, specifically analyzing how they relate to the managing data-as-a-product paradigm.

Information Plane

As we saw in *Chapter 2, Data Products*, a pure data product is the unit of modularization of the data architecture. Each pure data product aims to make one or more data assets accurate, relevant, reusable, and composable. Therefore, a cornerstone of this approach to data management is the idea of focusing only on relevant data, meaning data that can support real business cases and generate tangible value for the organization.

Pure data products have the fundamental responsibility of preserving the value of the data assets they manage, ensuring that these assets remain relevant and accurate over time for continued use. Since data increases in value the more it is used and integrated with other data, pure data products are also responsible for making the reuse and composability of the data assets they manage easier. For this reason, pure data products not only manage the data itself but also handle all related metadata. Metadata is an integral part of a pure data product, sharing the same life cycle of the managed data.

Due to the shift in metadata ownership to the product team and the metadata quality assurance policies implemented by the XOps platform, pure data products consistently offer the necessary context for users to understand and easily use the exposed data. This means that pure data products present not just raw data, but contextualized information through metadata. As a result, their zone of accountability extends beyond the data plane into the information plane (*Figure 11.4*).

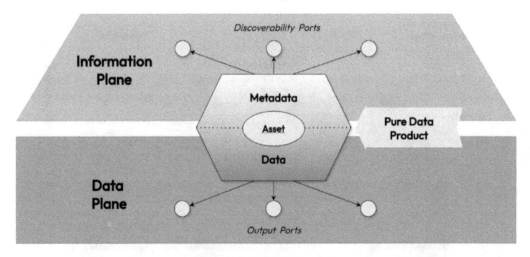

Figure 11.4 – Information plane

Unlike traditional approaches, a data product-centric approach assigns the sole responsibility for managing data and metadata—essentially the data plane and information plane within the information architecture—to the product team. Federated governance teams focus on defining which metadata should be collected and how it should be represented to enhance interoperability between data products and facilitate their consumption. However, these teams do not have any direct responsibility for managing the metadata itself.

Knowledge plane

In the previous chapter, we saw how the conceptual model created during the design of a data product can, once properly formalized, be used not only to support development but also to facilitate data consumption. The conceptual model offers a broader context for the exposed data asset, making it easier to use, reuse, and compose. While metadata helps us grasp the meaning of the exposed data, the conceptual model provides the domain knowledge necessary for interpreting and using that data effectively.

Unlike metadata, the conceptual model is not a part of the product and does not share its life cycle. It is simply referenced by the data product. Data products within the same subdomain (e.g., marketing and sales) can reference the same conceptual model, but none of them own it. The conceptual model is a shared artifact that evolves over time, driven by the needs of the data products while remaining independent of them.

Conceptual models for individual subdomains are often represented through **subdomain ontologies**, which are developed independently by federated modeling teams within each subdomain. This setup makes it easier to use and understand the exposed data within a subdomain. However, since these ontologies aren't connected to each other, they provide little support for using data outside of the subdomain or for combining exposed data from different subdomains.

Each subdomain, in fact, views the organization's reality from different perspectives and models it accordingly. While subdomain ontologies share many core concepts (such as customer, product, and contract), they represent these concepts in different ways based on their specific experiences, processes, and goals. As a result, when a consumer needs to use data from different subdomains, they must first understand each individual model. After that, they need to find a way to standardize the relevant concepts across these subdomains and normalize the data's semantic representation before they can effectively use it (*see Figure 11.5*).

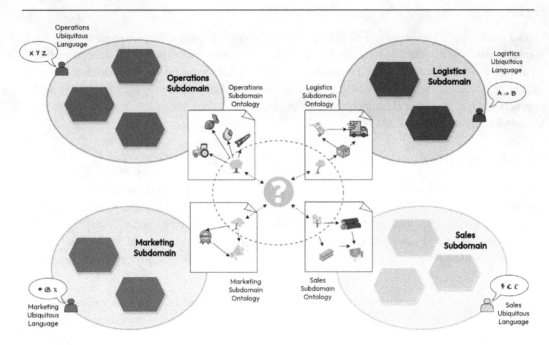

Figure 11.5 – Semantic gap between subdomains

This process of translation and semantic normalization would obviously be much more complex and time-consuming without the subdomain ontologies. Nevertheless, it remains a non-trivial task and is potentially prone to misunderstandings and translation errors. Additionally, this activity must be completely repeated each time a new consumer needs to access data from multiple subdomains.

It is therefore useful, in terms of knowledge management, not to stop at the formalization of subdomain ontologies but also invest resources in building a **domain ontology** that consolidates this semantic normalization work. A domain ontology contains all the core concepts for the organization that are common across multiple subdomains. The goal of a domain ontology is to provide a minimal definition of these concepts that can be accepted by all while allowing the subdomains to extend these definitions in a way that is functional to their specific operational contexts.

The domain ontology connects the ontologies developed by the various subdomains, ensuring a basic level of semantic interoperability among them. In other words, since every concept present in multiple subdomains can be traced directly or indirectly to a higher-level concept in the domain ontology, it becomes easier to determine if two concepts represent the same thing and what basic properties they share. This knowledge is essential for correlating data that instantiates the same concept across different subdomains. It provides the necessary information to make such data semantically interoperable and, therefore, composable (*Figure 11.6*).

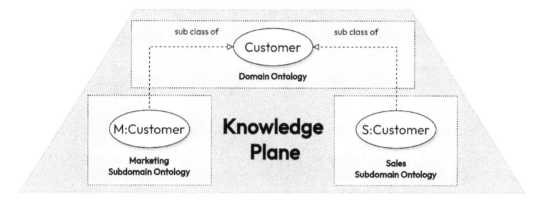

Figure 11.6 – Knowledge plane

Defining a domain ontology is much more complex than defining a subdomain ontology. When modeling an ontology within a subdomain, the people involved generally share similar mental models shaped by their common context. However, creating a domain ontology requires negotiation and agreement—often involving compromises—among individuals working in different contexts, each with their own unique perspectives on the organization.

Defining and evolving a subdomain ontology alongside data products adds extra costs to developing new business cases. Similarly, evolving the domain ontology incurs further costs. However, these upfront costs generate significant downstream value. Data must be contextualized to be usable, and the conceptual model provides this context. If not created upfront, it will need to be produced downstream by consumers, multiplying costs. Delaying knowledge modeling results in short-term savings but leads to greater negative impacts during data consumption (*Figure 11.7*).

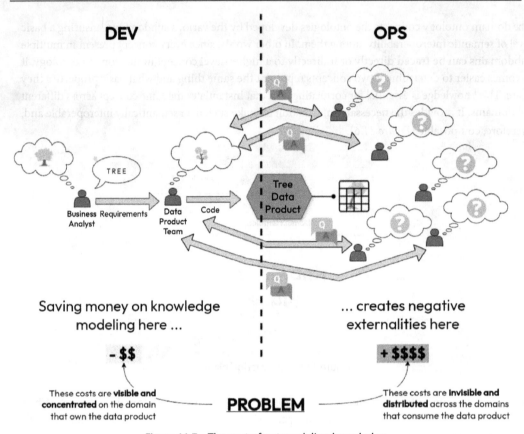

Figure 11.7 – The cost of not modeling knowledge

To avoid these negative impacts, it is therefore necessary to manage knowledge proactively rather than reactively. Let's see how in the next section.

Managing enterprise knowledge

To unlock the full potential of data, a deliberate, structured, and incremental approach to knowledge management is essential. But what does this process entail, and how can it be effectively achieved? In this section, we'll explore the modeling of knowledge through an enterprise ontology and the operational processes necessary to bring this model to life.

Enterprise ontology

In the previous section, we discussed the importance of establishing a common domain ontology across the organization to ensure the interoperability of models defined at the subdomain level and to prevent the creation of knowledge silos. The domain ontology, together with the more specific subdomain ontologies that extend it, constitutes what is commonly referred to as the **enterprise ontology** (*Figure 11.8*).

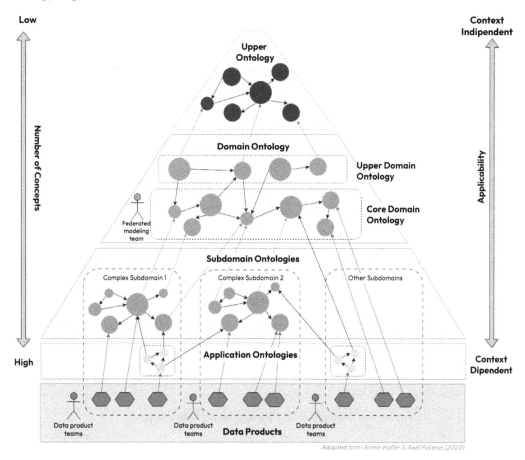

Figure 11.8 – Enterprise ontology

The **enterprise ontology** is a collection of related ontologies organized hierarchically by the level of abstraction of their concepts. Higher-level ontologies are less applicable for mapping real data but help maintain consistency and interoperability among lower-level ontologies by providing a shared vocabulary and structure. As we move down the hierarchy and create more specific ontologies, the concepts become increasingly specialized and better suited for mapping actual data generated by the organization.

Let's examine this hierarchical structure one level at a time.

Upper ontology

At the highest level of abstraction is the upper ontology, which defines fundamental concepts, such as object, person, time, event, and space. These concepts are universal and form the basis for creating more specific ontologies. Standard upper ontologies, such as **Basic Formal Ontology** (**BFO**) and GIST, provide established frameworks that make knowledge modeling easier in complex domains. Using a standard upper ontology, rather than creating a custom one or having none at all, usually improves the model's efficiency, interoperability, and quality. The following figure shows some of the key concepts and relationships defined within GIST upper ontology (*Figure 11.9*).

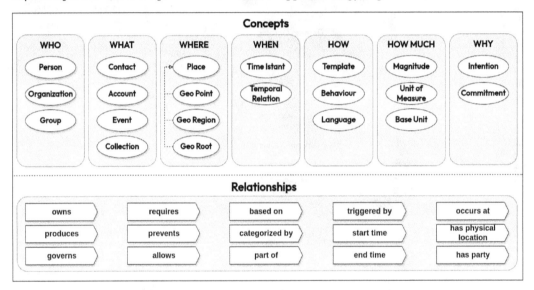

Figure 11.9 – Sample of GIST's concepts and relationships

Domain ontology

Below the upper ontology is the **domain ontology**, which defines the concepts and relationships specific to a business domain (e.g., finance, insurance, retail) or a field of knowledge (e.g., economics, biology, engineering). It serves as an intermediate layer between the general upper ontology and the more specialized subdomain ontologies.

A domain ontology can optionally be divided into the following two sub-levels:

- **Upper domain ontology**: This contains all the key concepts and relationships for a specific domain at a level of abstraction that allows them to be used across different organizations operating within that domain. It does not include specialized elements unique to any particular organization. For example, in the upper domain ontology of a hypothetical food company, the concept of Product from the upper ontology could be extended to define the concept of Perishable Product.

- **Core domain ontology**: This extends the concepts and relationships from the upper ontology or upper domain ontology, adding elements that characterize a specific organization. For instance, Food Corp might need to extend the concept of `Perishable Product` by adding the concept of `Weather-Sensitive Product`, as a significant part of its specific production and sales processes rely on weather forecasts.

In general, it is beneficial to have this division of the domain ontology when aiming to ensure interoperability not only within the organization but also potentially with external entities. In such cases, a standard domain ontology or a portion of it can be chosen as the upper domain ontology. Its concepts can then be further specialized in the core ontology to include knowledge that is meaningful only to the specific organization. The presence of a standard upper domain ontology ensures interoperability with other organizations operating within the same domain that adopt it.

Subdomain ontologies

Below the domain ontology, subdomain ontologies are defined by extension. Each **subdomain ontology** specifies the concepts and relationships that characterize a specific subdomain within the organization (e.g., marketing, sales, operations, logistics). Not all subdomains need to extend the enterprise ontology; this extension occurs only when there is a genuine need to specialize certain concepts.

For instance, the operations and logistics subdomains may need to extend the concept of `Weather-Sensitive Product` to add relevant properties or relationships for production (e.g., `bill of materials`) or shipping (e.g., `shipment unit type`). In contrast, the sales and marketing subdomains might not require any extensions, as the existing concepts and relationships in the enterprise ontology are adequate for their modeling needs.

Finally, within subdomain ontologies, application ontologies can be defined. An **application ontology** is a specialized ontology designed for a specific use or application. It references concepts from higher-level ontologies or other subdomain ontologies. Application ontologies are particularly useful for creating visualizations and data views tailored to specific business cases. For example, the logistics subdomain might define the concept of `Monthly Product Stock Report` in an application ontology to describe a data view defined for BI tools that integrates product, stock, and time-related data. While it's not always necessary to create a specific concept for these application views—since the integrated data often connects to existing concepts in higher-level ontologies—doing so can help, in some cases, applications better understand the data views prepared for them.

The transition from the domain ontology to the subdomain ontologies marks the shift from knowledge shared across the entire organization to the specific knowledge of various subdomains. An enterprise ontology is not a single model that unifies and normalizes all of the organization's knowledge. Instead, it is a collection of models defined through ontologies, where the same business entities may be represented differently in each subdomain. However, these different representations are still linked through the central domain ontology that ensures their interoperability at a semantic level. The following figure illustrates this hierarchical and federated structure in the context of modeling the product entity (*Figure 11.10*).

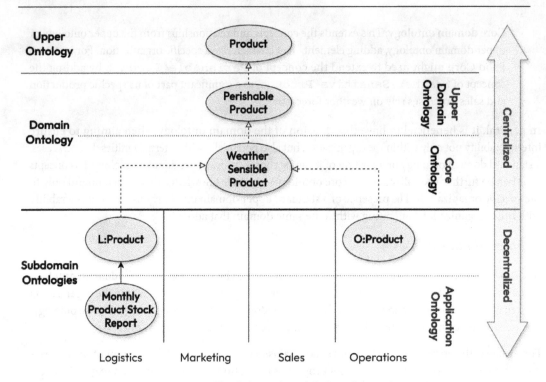

Figure 11.10 – The federated model of product entity

Federated modeling team

Each subdomain can independently decide if and how to define a subdomain ontology. The responsibility for defining subdomain ontologies is therefore distributed across the individual subdomains. On the other hand, the responsibility for defining the domain ontology must be centralized in one or more **modeling teams** in order to achieve the definition of a shared model that is common to the entire organization. These teams are specific governance teams, organized according to a federated or hybrid model.

A modeling team typically includes one member from each relevant subdomain, who dedicates part of their time to the federated team's work. Members of a federated modeling team must have specific business knowledge related to the area being modeled. They must also have the authority and mandate to negotiate and reach agreements with other team members on the shared model that needs to be defined and then adopted by each subdomain.

Since the federated modeling team consists of businesspeople, it's necessary to complement them with modeling experts (e.g., information architects, data stewards, and so on) who can assist in building and formalizing the model. These more technical individuals can either be dedicated to the federated modeling team or part of a **complicated subsystem** team that collaborates with it. The first approach is generally useful when the modeling area is in its initial definition phase, as dedicated technical resources improve interactions without causing inefficiencies. However, when the area is more developed and the team's activities are less frequent, the second approach is preferable to maintain quality while optimizing resource allocation.

In complex organizations, it's often necessary to establish multiple modeling teams focused on different areas that need to be modeled. A single team of 10-15 people is unlikely to possess all the knowledge required to model the entire organization effectively. Therefore, it's important to break down the business knowledge into smaller, more manageable parts that dedicated teams can focus on and model effectively.

Within the broader scope of organizational knowledge to be modeled, we define a **knowledge domain** as a coherent subset assigned to a modeling team for autonomous modeling. Some organizations refer to these as **data domains** to distinguish them from business subdomains. However, we prefer the term knowledge domain as it more accurately reflects the level within the information architecture where these domains are defined.

The knowledge domains are orthogonal to the business subdomains, as they aim to identify a portion of the business knowledge that can then be represented through a model shared by all the business subdomains.

Subdomains divide the organization into horizontal slices based on the capabilities of each function, department, or unit. In contrast, knowledge domains categorize the organization according to business entities (e.g., customer, product, order) or business processes (e.g., procure-to-pay, order-to-cash) that cut across the organizational structure (*Figure 11.11*).

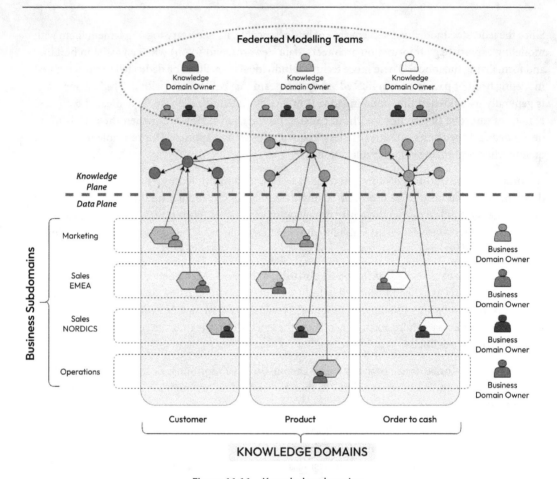

Figure 11.11 – Knowledge domains

In organizations where multiple federated modeling teams are necessary, it is also essential to establish a governance and coordination framework to ensure that each ontology produced meets the same quality standards and can easily integrate with others to form a cohesive enterprise ontology. In such cases, the shared rules for governing and coordinating the development of the enterprise ontology can be managed informally through collaboration among the various modeling teams, depending on their needs, or formalized through the establishment of a dedicated committee.

Managing knowledge as a product

Each ontology within the domain ontology is managed by a dedicated modeling team responsible for its definition, development, and evolution. Like data products, the development of each ontology must always be driven by specific business needs. There's no reason to model a particular area of the organization unless there are relevant use cases that require it. Ontologies only have value if they are actively used. Once an ontology is defined, it continues to evolve over time alongside the business cases that rely on it.

Since ontologies have clear ownership, a dedicated team, and evolve over time to meet the needs of their users, it makes sense to manage them as products rather than projects. In this context, where knowledge is managed as a product, federated modeling teams can be seen as **stream-aligned teams** delivering **knowledge products** to consumers in an **X-as-a-service** mode. These teams work closely with the teams responsible for data products to incrementally and synergistically build a robust information architecture, guided by use cases, that maximizes the potential value of the managed data (*Figure 11.12*).

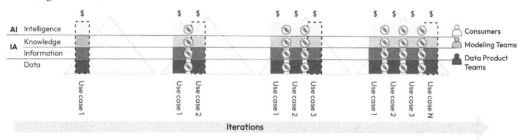

Figure 11.12 – Building the information architecture one product at a time

Just like with data products, it makes sense to develop also for knowledge products a **self-serve platform** to support their development and life cycle, reducing the cognitive load on modeling teams. This platform should go beyond simply cataloging and making ontologies accessible. It should also provide tools for their development and operation (i.e., **KnowledgeOps**) as well as for their validation based on defined governance policies (i.e., **computational policies**).

This knowledge product development platform is logically part of the XOps platform, but it can be implemented using different tools than those used for data products. Since data and knowledge product development work together and both contribute to the information architecture, it's crucial that the tools and services supporting data products are well integrated with those for knowledge products.

We will refer to the knowledge management approach described in this section as **knowledge mesh**. This approach, particularly in organizations that need multiple modeling teams, follows the same four founding principles of data mesh, but applies them to the knowledge plane instead of the data plane (*Figure 11.13*).

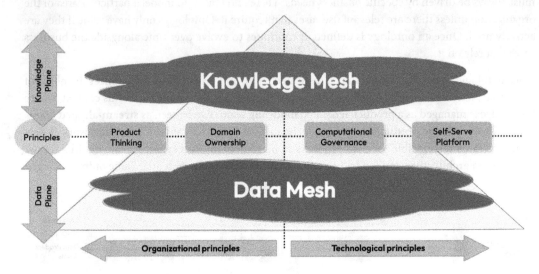

Figure 11.13 – Knowledge mesh

Building an enterprise knowledge graph

So far, we have seen how to define and build an enterprise ontology. Now, let's explore how to connect it to the developed data products in order to create an enterprise knowledge graph that can represent the entire information architecture in one unified model.

Knowledge Graph Architectures

A knowledge graph is created when the conceptual model, represented by one or more interconnected ontologies, is linked to data that instantiates the concepts within the model. There are three main types of knowledge graph architectures, based on how and when the connections between concepts and data are established (*Figure 11.14*).

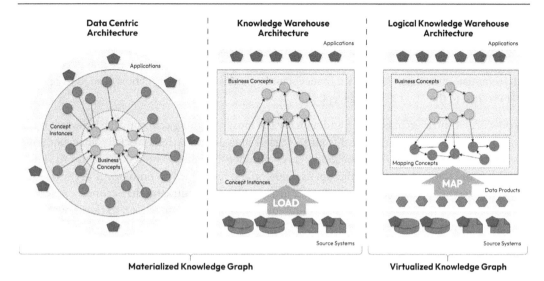

Figure 11.14 – Knowledge Graph Architectures

The first two architectures shown in the figure are examples of **materialized knowledge graphs**. In these graphs, the data that represent various business concepts are directly included as concepts within the ontology. For example, if the ontology defines the concept of `Customer` (**business concept**) and Andrea Gioia is a customer (**factual data**), then Andrea Gioia is represented directly in the ontology as a concept that instantiates the concept of `Customer`.

The architectures of materialized knowledge graphs differ based on the role the knowledge graph plays within the broader enterprise data management architecture. Specifically, there are two main types of materialized knowledge graphs:

- **Data-centric architecture**: In this architecture, data exists only within the ontology. Applications access the data through the ontology for both reading and writing, without maintaining a local copy. This is the knowledge graph architecture used in the **data-centric approach** to data management discussed in *Chapter 1, From Data as a Byproduct to Data as a Product*.

- **Knowledge warehouse architecture**: In this architecture, data is copied from the applications that manage them into the ontology. The knowledge graph is used to read the integrated data but not to write it back. In this case, ownership of the data remains with the applications that create and manage it.

The third type of architecture, known as the **logical knowledge warehouse architecture**, is based on a **virtual knowledge graph**. In a virtual knowledge graph, the data that instantiates the concepts are not directly represented in the ontology; instead, they are simply referenced. The ontology includes technical concepts that describe how business concepts map to the data sources containing the actual instances.

As a result, even though the data isn't directly included in the ontology, it still holds all the necessary information to retrieve it from the source systems. For example, the customer Andrea Gioia may not be directly represented in the ontology, but it includes the concept of `Customer Data Product`, which instantiates the `Data Product` concept and contains all the information needed to access the data related to the `Customer` concept.

The knowledge-centric architecture is undoubtedly the most intriguing conceptually. However, it is also the most challenging to implement from an organizational perspective, especially since modern IT architectures are still heavily application-centered rather than data-centered. It's better to view this architecture not as a starting point but as a potential target. Once an ontology is defined and data can be accessed through it, envisioning a transition to a data-centric model becomes easier. In this model, new applications gradually start to not only read data but also write it to the knowledge graph, all without storing a local copy.

In this chapter, we will delve into knowledge warehouse architectures, both logical and physical (i.e., non-logical). These architectures are easier to implement in today's IT landscape. They also function similarly by providing a **semantic layer** that facilitates the use of physical data. From the consumer's perspective, they are indistinguishable; whether the data is materialized within the ontology or simply referenced is just a technical detail.

We will specifically focus on the logical knowledge warehouse architecture, as it allows referenced data to be materialized within the ontology at any time, while the reverse is not possible. In the next section, we will look at how to connect the data exposed by data products to the concepts within the ontology.

Connecting data and knowledge plane

A knowledge graph elegantly represents the entire information architecture within a single model. It provides data consumers with a centralized access point for data, information, and knowledge. This allows them to use the same tools to search within the knowledge and information plane first (i.e., search for data) and then to search in the data plane (i.e., search in data).

To create a knowledge graph, it is essential to connect the data in the data plane with the concepts in the knowledge plane using the metadata from the information plane. The first step involves specifying where and how data products can define within their descriptor files this **semantic link** between the data they expose and the concepts in the domain or enterprise ontology.

In the **Data Product Descriptor Specification (DPDS)**, all metadata related to the data exposed by data products are defined using standard annotations applied to the same schemas used to describe the structure of data. Within a schema—regardless of its specific format (e.g., JSON Schema, AVRO Schema, Protobuf, etc.)—the standard annotations from DPDS can be associated with a specific entity, or with a specific field within an entity. The following example illustrates how the `kind` annotation is used in a JSON Schema to add metadata to the structure of a hypothetical dataset containing movies:

```
{
    "$schema": "http://json-schema.org/draft-07/schema#",
```

```
      "title": "Simplified Movie Object (Compact)",
      "type": "object",
      "kind": "tabular",
      "properties": {
        "movieId": {
          "kind": "identifier",
          "type": "string"
        },
        "name": {
          "kind": "attribute",
          "type": "string"
        }
      }
    }
  }
```

The semantic link between the data and the concepts they instantiate is made in DPDS using the
s-context annotation defined at the entity level in the data schema, as shown in the following example:

```
{
    "$schema": "http://json-schema.org/draft-07/schema#",
    "title": "Simplified Movie Object (Compact)",
    "type": "object",
    "kind": "tabular",
    "s-context": {
        "s-base": "https://schema.org",
        "s-type": "[Movie]",
        "movieId": null,
        "title": "name"
    },
    "properties": { [...]
```

In this case, the semantic link indicates that the data in the dataset are instances of the Movie
concept from the schema.org ontology. The movieId field does not correspond to any attribute
in the Movie concept, while the title field corresponds to the name attribute. The s-context
annotation also allows for semantic links to properties of different concepts, meaning there's no need
for a strict one-to-one relationship between a dataset and a concept. This is useful when data related
to different concepts are denormalized into a single dataset.

For example, in schema.org, Movie is connected to the Person concept through the director
relation, and Person is connected to the Country concept via the nationality relation. If the
dataset includes a field called directorCountryName, it could be linked to the ontology in the
s-context annotation like this:

```
"directorCountryName":
  "director[Person].nationality[Country].name"
```

By proactively including this semantic linking information, data products are immediately ready to be integrated into the virtual knowledge graph upon publication. Let's see how.

Knowledge-driven data management

Using the information in the descriptor file, the XOps platform can record the connections between the data exposed by published data products and the concepts in the enterprise ontology. This allows consumers to navigate the information architecture either from the top down or from the bottom up, depending on their needs.

It's particularly beneficial for the XOps platform to represent not only business concepts but also all the metadata it collects through an ontology for the following reasons:

- Unifying the search experience by providing a single service and a common query language (e.g., SPARQL) to search across business concepts and metadata

- Providing a unified representation encoded according to shared standards (e.g., RDF) of all information architecture

To accomplish this, it's essential to define one or more ontologies that define concepts for each type of metadata being collected (e.g., data product, output port, dataset). The actual metadata gathered should then be represented as instances of these concepts. We will refer to this collection of ontologies used to map the managed metadata as the **metadata ontology**. When combined with the instances of the collected metadata, the metadata ontology forms the **metadata knowledge graph**. This knowledge graph is a part of the unified model we aim to create to represent the entire information architecture, which is referred to as the **enterprise knowledge graph**.

Finally, it's important to define within a **mapping ontology** all the concepts necessary to link the data exposed by the data products with the associated metadata and with the business concepts they represent. Since mappings are also metadata, the mapping ontology can be seen as part of the metadata ontology. However, because it specifically focuses on representing relationships between data, metadata, and business concepts, we prefer to treat it as a separate, standalone ontology (*Figure 11.15*).

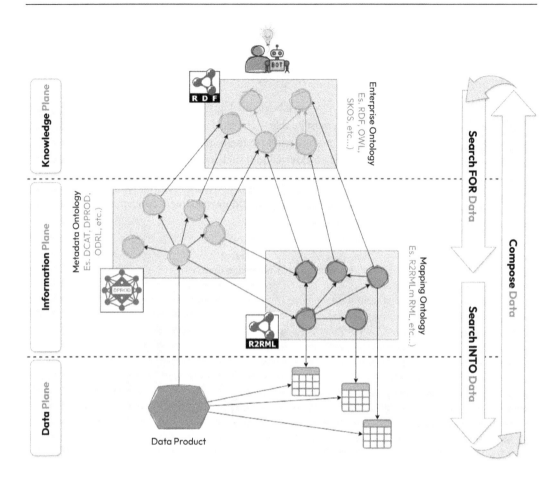

Figure 11.15 – Enterprise knowledge graph

Although it is possible to define these ontologies in a custom way, it is preferable to use standard ontologies and extend them where necessary. Using standard ontologies ensures the following:

- Interoperability with external tools and organizations
- Greater ease of understanding for AI agents, as these ontologies are part of their training dataset, leading to improved accuracy and efficiency in their tasks

Data Product Ontology (**DPROD**) is an example of a standard ontology designed to describe data products and their key associated concepts. It can be complemented by other more specific ontologies to represent various types of metadata. For instance, **Open Digital Rights Language** (**ODRL**) can be used to outline data usage policies, **provenance ontology** (**PROV-O**) can be used to track data lineage, and **Data Quality Vocabulary** (**DQV**) can be used to define data quality metrics.

Otherwise, **RDB to RDF Mapping Language** (**R2RML**) and **RDF Mapping Language** (**RML**) are examples of standard ontologies for mapping business concepts to physical data instances. R2RML is specifically used for mapping structured data stored in relational databases, while RML is utilized for mapping semi-structured data.

The metadata ontology and mapping ontology must be defined upfront based on the metadata that is intended to be collected. The XOps platform must be able to convert the collected metadata into instances of the concepts defined in the metadata ontology and save them in the enterprise knowledge graph, which can be managed directly or by using an external tool such as a graph database or a triplestore.

The domain ontology, on the other hand, is built incrementally over time based on the needs of specific use cases (*Figure 11.16*). In this case, the modeling teams add new business concepts directly to the enterprise knowledge graph as they define them.

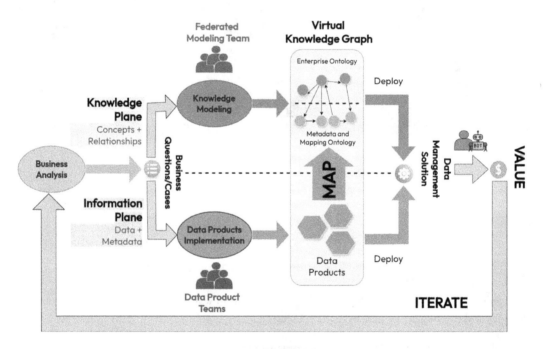

Figure 11.16 – Enterprise knowledge graph development process

Leveraging modern AI

At the top of the information architecture lies intelligence, defined in this context as the ability to leverage data, information, and knowledge to make informed decisions and take meaningful actions that create value for the organization. Human intelligence is nowadays increasingly being complemented, integrated, and, in some cases, even replaced by AI. In this section, we will explore how modern AI techniques, as well as those that will likely emerge in the near future, can benefit from a robust and comprehensive information architecture.

The generative AI revolution

Generative AI has undoubtedly elevated AI systems to a level that was unimaginable just a few years ago. The range of results and potential applications is vast, and the industry is investing in these technologies at an unprecedented rate.

On one hand, generative AI models continue to improve month after month, pushing the limits of what they can achieve, particularly in generating high-quality text, images, and videos. On the other hand, several significant challenges remain that hinder their widespread adoption within organizations. Three key issues stand out:

- **Lack of specialized knowledge**: Generative AI models often lack the deep, specialized knowledge necessary to fully understand how your organization operates and what sets it apart from competitors. Trained on large, generic datasets, these models can produce quality content, but they struggle to grasp the nuances and technical details that are unique to specific domains. This gap in understanding poses a challenge for adopting generative AI in domain-specific business cases.

- **Unpredictable output**: Generative AI models can produce inconsistent and sometimes unreliable results, especially for complex or nuanced tasks. Because these models are probabilistic, small differences in input can lead to widely different outcomes, making it hard for organizations to fully trust the AI. This unpredictability poses challenges in areas such as customer communications, legal documents, or marketing materials, where errors or inaccuracies can have significant consequences.

- **Limited reasoning skills**: Generative AI models have restricted reasoning capabilities, struggling to draw logical connections between disparate pieces of information. While they excel in recognizing patterns, they often falter at tasks requiring deeper cognitive processes, such as making inferences, decision-making, or understanding nuanced cause-and-effect relationships. This limitation diminishes their effectiveness in addressing complex business challenges that demand strategic reasoning and planning.

Implementing a robust information architecture can help mitigate these issues and enhance the number and value of business cases where generative AI can be successfully deployed. Let's explore how this can be done.

Boosting generative AI with domain knowledge

Generative AI models don't naturally possess the specific knowledge needed to work effectively within a particular organizational context. This is because the necessary information isn't publicly available and therefore isn't included in their training set. As a result, the only way to use a generative AI model to tackle business cases requiring specialized knowledge—without resorting to expensive full or partial retraining of the model—is to provide the relevant contextual information along with the specific request. This provided context is essential for the model to respond accurately.

Retrieval-augmented generation (**RAG**) is a widely used technique to retrieve relevant data or documents from organizational systems and pass them to the model to perform a domain-specific task. The data provided to the model during inference gives it the context needed to execute the task in a more accurate way. However, since this information is in the form of structured or unstructured data, the model must perform the same cognitive effort as a human to understand the meaning of the provided data and how to use it effectively to answer the received request.

GraphRAG extends the vanilla RAG technique by passing not only raw data but also relevant concepts and relationships from the enterprise ontology to the generative AI model. This enables the model to reason more effectively, as it no longer needs to derive a conceptual model from the data—it already has access to it. This access is key to improving the accuracy of the model. It allows the model to connect related concepts, resulting in answers that reflect a deeper understanding of the context. GraphRAG not only addresses the lack of specific domain knowledge in generative AI models but also reduces unpredictability in responses and enhances reasoning capabilities, leading to more insightful and context-aware answers.

In contrast to vanilla RAG, GraphRAG cannot rely solely on available data; it also requires an ontological model of the domain in which it operates. This model can be inferred from data using specific AI techniques, but an automatically generated conceptual model will never match the quality of one intentionally defined by domain experts.

To fully leverage GraphRAG, it is crucial to have a robust information architecture that manages both data and knowledge, ensuring these two planes are properly connected. This enables the retrieval techniques used by GraphRAG algorithms to navigate the entire information architecture, seamlessly searching through concepts and data to provide the best possible context for the generative AI model to perform effectively.

Future-proof your AI investment

Having a solid information architecture allows organizations to fully harness the potential of generative AI today while ensuring that their AI investments remain relevant in the future.

Research in AI is increasingly focused on developing new hybrid architectures that combine statistical techniques, such as those used in neural networks, with traditional symbolic AI methods based on knowledge graphs. This integration aims to overcome the limitations of both approaches. This new way of thinking about AI is known as **neuro-symbolic AI**, which draws inspiration from the way the human brain functions, as described by Daniel Kahneman in his book, *Thinking, Fast and Slow*.

Kahneman identifies two systems of thought: System 1 and System 2 (*Figure 11.17*).

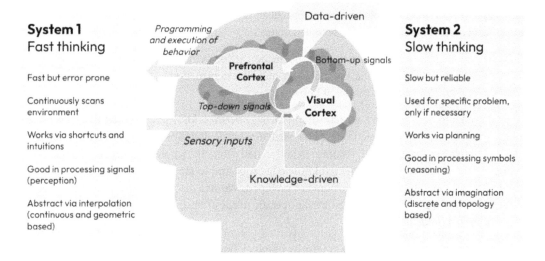

System 1
Fast thinking

Fast but error prone

Continuously scans
environment

Works via shortcuts and
intuitions

Good in processing signals
(perception)

Abstract via interpolation
(continuous and geometric
based)

Data-driven

Programming
and execution of
behavior

Prefrontal
Cortex

Bottom-up signals

Top-down signals

Visual
Cortex

Sensory inputs

Knowledge-driven

System 2
Slow thinking

Slow but reliable

Used for specific problem,
only if necessary

Works via planning

Good in processing symbols
(reasoning)

Abstract via imagination
(discrete and topology
based)

Figure 11.17 – Systems of thought

System 1 is quick, intuitive, and automatic, handling tasks with little effort or conscious thought. On the other hand, System 2 is slower and more analytical, used for solving complex problems that require careful reasoning. Neuro-symbolic AI reflects this duality by combining the fast, efficient generation of insights from neural networks (such as System 1) with the structured, logical reasoning capabilities of symbolic AI (such as System 2). This combination results in a more balanced and capable AI system that not only generates relevant content but also interprets it in meaningful ways.

However, even as AI continues to evolve and incorporate more neural networks and knowledge models, it cannot automatically generate the unique knowledge that only an organization possesses. Future AI systems will likely become so advanced that they render current techniques, such as RAG and GraphRAG, obsolete, but they will still need to acquire contextualized knowledge in one form or another to function effectively within a specific organization. This principle also applies to people, the highest form of intelligence we know so far. No matter how intelligent and skilled a professional is, he or she will need time to understand the nuances of a new organization before becoming fully productive. The same is true for both current and future AI systems, regardless of their level of intelligence.

Therefore, organizations that invest in formalizing their specific domain knowledge will be better positioned to take full advantage of the innovations offered by AI, enabling them to respond more quickly and effectively to new opportunities.

Summary

In this chapter, we explored how to create a comprehensive information architecture that builds upon managed data assets to maximize their potential value.

We initially examined the different layers of an information architecture, analyzing how data transforms into information through the addition of metadata, and how information, organized into interrelated concepts, evolves into an explicit model of knowledge.

We then delved into how to collect, structure, and manage business knowledge through an enterprise ontology. Specifically, we examined in detail the structure of the domain ontology, which, unlike the subdomain ontologies that extend it, is common and shared across the whole organization. Its purpose is to prevent the creation of disconnected knowledge silos and to lay the foundation to guarantee the semantic interoperability of data products developed. We also discussed how its definition and incremental evolution over time, guided by the specific needs of business cases, is managed by one or more federated modeling teams.

Finally, we explored how to connect the enterprise ontology (knowledge plane) with the data products (data plane) using semantic links defined by the data product teams within descriptor files. These links create connections between the data exposed by the data products and the concepts in the enterprise ontology, resulting in a unified model of the entire information architecture known as the enterprise knowledge graph. Consumers can use the knowledge graph as a single model to search for relevant data and navigate within it. The consumers benefiting from the contextual information provided by the knowledge graph include both people and AI agents. In the final section of the chapter, we illustrated how a comprehensive information architecture enables these AI agents to effectively implement domain-specific business cases with high accuracy.

Further reading

For more information on the topics covered in this chapter, please see the following resources:

- *The Economy's Power Shift*. P.F. Drucker in Harvard Business Review (1992). `https://hbr.org/`

- *Conceptual Framework for Financial Reporting*. Accounting Standards Board (1989). `https://www.ifrs.org/issued-standards/list-of-standards/conceptual-framework/`

- *Data Product Ontology (DPROD)*. OMG (2024). `https://ekgf.github.io/dprod/`

- *R2RML: RDB to RDF Mapping Language*. W3C (2012). `https://www.w3.org/TR/r2rml/`

- *Introducing the Knowledge Graph: things, not strings*. A. Singhal (2012). `https://blog.google/products/search/introducing-knowledge-graph-things-not/`

- *The GraphRAG Manifesto: Adding Knowledge to GenAI*. Neo4J (2024). `https://neo4j.com/blog/graphrag-manifesto/`

- *A Benchmark to Understand the Role of Knowledge Graphs on Large Language Model's Accuracy for Question Answering on Enterprise SQL Databases*. J. Sequeda, D. Allemang, B. Jacob (2023). `https://arxiv.org/abs/2311.07509`

- *Increasing the LLM Accuracy for Question Answering: Ontologies to the Rescue!*. D. Allemang, J. Sequeda (2024). `https://arxiv.org/abs/2405.11706`

- *Thinking, Fast and Slow*. Daniel Kahneman (2013). `https://a.co/d/cYUteu5`

- *The Rise of Cognitive AI*. G. Singer (2021). `https://towardsdatascience.com/the-rise-of-cognitive-ai-a29d2b724ccc`

- *Thrill-K: A Blueprint for The Next Generation of Machine Intelligence*. G. Singer (2021). `https://towardsdatascience.com/thrill-k-a-blueprint-for-the-next-generation-of-machine-intelligence-7ddacddfa0fe`

Get This Book's PDF Version and Exclusive Extras

Scan the QR code (or go to `packtpub.com/unlock`). Search for this book by name, confirm the edition, and then follow the steps on the page.

Note: Keep your invoice handy. Purchases made directly from Packt don't require one.

12

Bringing It All Together

Here we are at the end of this book, which I hope will also be the beginning of your journey into adopting the managing data-as-a-product paradigm. Throughout the chapters, we explored what it means to manage data as a product—from understanding the core reasons driving this new approach to diving into technical implementation details and, finally, discussing the organizational shifts needed to make it work. Along the way, I hope you've picked up some insights and practical tips to help you make thoughtful decisions and avoid critical mistakes that could hinder successful adoption.

In this final chapter, before passing the ball to you, I'd like to summarize some recurring themes in this book and, more broadly, share my perspective on data management through three core beliefs and a few practical suggestions. I hope these will spark reflection and serve as a helpful compass as you navigate the fascinating world of data management.

This chapter will cover the following main topics:

- Core beliefs shaping the future of data management
- Setting yourself up for success
- Final remarks

Core beliefs shaping the future of data management

Through my experience in data management, I've developed three core beliefs that guide my work and inspired each chapter of this book. These beliefs focus on how and why data management is evolving. In this section, I'll briefly outline them.

Data management is not a supporting function anymore

Over the past 40 years, the world we live in has changed profoundly, becoming progressively more **volatile, uncertain, complex, and ambiguous** (**VUCA**). Globalization, the digital revolution, natural disasters linked to climate change, and an increasingly unstable geopolitical landscape have contributed to the environment in which organizations operate in more hostile ways, and where it is difficult not only to predict what will happen but also to understand the potential impacts of what is already happening. These profound transformations have significantly changed the way organizations compete in the market, forcing them to deeply rethink their strategies in order to survive and thrive in a constantly evolving environment. With each shift in competitive advantage theories, the role of data within firms has also evolved, gradually moving, at least theoretically, from being a useful but non-strategic resource to becoming today, and likely even more in the future, one of the most important productive factors for any type of business (*Figure 12.1*).

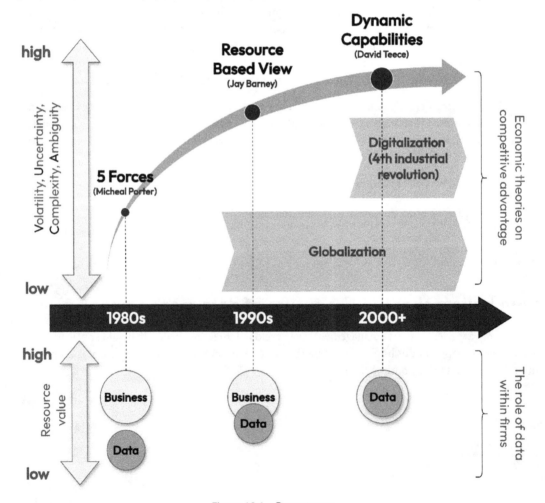

Figure 12.1 – Data at core

In the early 1980s, the world was relatively stable compared to today, and companies operated in an environment where competitive dynamics were more predictable. During this time, Michael Porter developed his well-known theory of competitive advantage. He argued that a company can achieve a competitive advantage by strategically selecting the industry in which it operates and determining its approach within that industry based on five key external forces: buyer power, supplier power, rivalry among competitors, the threat of new entrants, and the risk that the product or service offered could be replaced by other products or services. In summary, for Porter, a company achieves a competitive advantage when it invests to build a product or service that is unique in terms of price and/or functionalities in a market whose structure (i.e., demand and supply dynamics, entry barriers, etc.) makes it difficult, once a leadership position is reached, to lose it. Therefore, according to Porter's **Five Forces theory**, described in his *Competitive Advantage: Creating and Sustaining Superior Performance* book, a company must invest in its early stages to carefully identify the market in which to operate, define the product or service to offer to that market, and implement it in a way that significantly differentiates the offering from competitors. The assumption behind this theory is that once a company establishes a competitive advantage through a product or service that is hard to imitate or replace, it can reasonably expect to benefit from it for an extended period. In the competitive environment that Porter's theory focuses on, data is mainly used to optimize scaling and exploitation processes. However, data itself isn't central to the business, and the role of data management is more supportive than critical in shaping strategy.

As globalization began to take hold, the world became increasingly more complex and unpredictable. The idea of relying solely on market analysis to create a unique offering was no longer sufficient. The growing complexity blurred the lines between different markets, while increasing uncertainty and volatility made it challenging to analyze the defining forces that shape these markets. Consequently, theories of competitive advantage started to shift their focus more toward internal organizational factors rather than external market forces.

In the *Resource-Based Theory: Creating and Sustaining Competitive Advantage* book published in 1990, Jay Barney suggests that the true competitive advantage of organizations mainly stems from the unique resources and capabilities they possess—those that are **valuable, rare, inimitable, and non-substitutable (VRIN)**. In a changing and uncertain market, an organization may need to adapt its offering multiple times, but thanks to its unique resources and capabilities, it can remain competitive over time. In the competitive environment that Barney's theory focuses on, data is one of the unique resources that an organization possesses and must invest in to build a competitive advantage. The data management function thus becomes more aligned with the business and central to strategy definition.

With the advent of the digital revolution (i.e., the *fourth industrial revolution*), VUCA have become defining elements of every market, rather than exceptions. During globalization, the focus of competitive advantage theories shifted from building a unique offering to managing unique resources. Now, in the digital era, the focus has shifted again—to the ability to continuously adapt, integrate, and reconfigure internal and external resources to develop new capabilities and quickly respond to market changes. In the *Dynamic Capabilities and Strategic Management: Organizing for Innovation and Growth* book, published in the early 2000s, David Teece formalizes this view. It suggests that an organization's competitive advantage depends on its ability to rapidly adapt to changes in its environment, which in turn relies on the following three key capabilities:

- **Sensing capability**: The ability to identify opportunities and threats in the environment in which the organization operates

- **Seizing capability**: The ability to mobilize available resources to capitalize on these opportunities and respond to identified threats

- **Transforming capability**: The ability to reconfigure mobilized resources to develop the necessary capabilities to adapt to changes

The implementation of these key capabilities, essential for organizational competitiveness, is driven by data. In this competitive context, data is no longer just a tool to optimize processes or a key resource for differentiating from competitors. Data has become a productive factor on which the activities of organizations are based, regardless of the market in which they operate. Thus, the data management function becomes an integral part of the business, as central as natural resources, labor, capital, and entrepreneurship.

For these reasons, I believe it's crucial to rethink how we approach data management and take action to make it more strategic. There's still a significant gap between the ideal role data management should play and its current role in most organizations. This book is my humble contribution to bridging this gap and supporting organizations as they rethink how they manage their data more strategically.

Data management is not just about data

Data management goes beyond the mere collection, storage, and sharing of data. Data only holds value when it is leveraged to create tangible benefits for the organization. As we explored in the previous chapter, effective data management must transcend the data itself and adopt a holistic approach to managing the entire information architecture.

Therefore, to manage data effectively, it's crucial to also manage metadata that transforms it into comprehensible information, along with the domain knowledge that allows us to utilize this information meaningfully within the organization. These activities—moving beyond simple collection, storage, and sharing—demand a rethinking of not just the technological architecture but also the organizational model (see *Figure 12.2*).

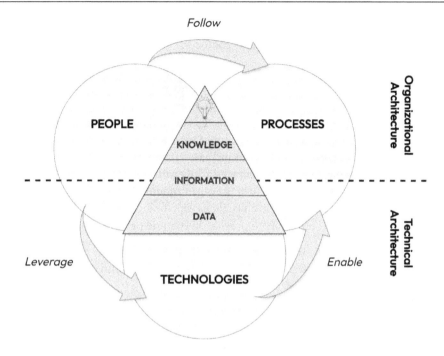

Figure 12.2 – Data management beyond just managing data

In *Chapter 1, From Data as a Byproduct to Data as a Product*, we identified tunnel vision as a primary cause of failure in data management solutions. This narrow focus often leads to an overemphasis on data and technology, causing us to lose sight of the broader system. People and processes are equally important in data management, especially when transitioning from handling raw data to managing an entire information architecture. It is the management of people, their interactions, and the ways these elements influence technology choices—and vice versa— that makes a data management solution a complex socio-technical system.

For this reason, I believe it is essential to adopt a systemic approach to building a data management solution that equally considers people, processes, and technologies, co-evolving these system elements in a synergistic manner. In this book, in presenting the managing data as a product paradigm, I have tried to follow this approach, offering a systemic view in both the description of problems and potential solution proposals, aimed at highlighting the socio-technical implications of each topic addressed.

Data management is not just an IT responsibility

Traditionally, data management is seen as a purely technical issue and therefore a responsibility of the IT department. However, this approach to data management is problematic in the VUCA world in which we live, as it creates ongoing tensions between IT, which manages the data, and the business units, which need to use this data to respond effectively to the challenges posed by an ever-changing competitive landscape. This internal struggle is not a contingent issue but a systemic one, caused by a resource—data—shared between distinct groups within an organization—IT and business units—that operate under different priorities and often pursue conflicting goals.

According to Ashby's *Law of Requisite Variety*, outlined in the *An Introduction to Cybernetics* book, for a system to efficiently manage and adapt to the complexity of its environment, it must possess a sufficient level of internal variability that reflects that of its environment. In other words, a system's ability to adapt and survive in response to external challenges depends on its internal capacity to mirror the complexity of the challenges; a simple system cannot solve a complex problem. For this reason, as the complexity of the competitive environment increases, organizations are also becoming more complex. Specifically, the business units that interact directly with the external environment need to create solutions faster, increasing the variability in the organization's output to stay competitive. This operational mode, which characterizes business units, aligns with the logic of a **differentiation economy**, where value is produced based on the ability to provide diverse and personalized solutions quickly.

On the other hand, IT operates in a **scale economy**, where value is created by optimizing the use of shared resources. To achieve this, IT tries to limit the variability of the resources it manages, reducing costs in both acquisition and operational management.

For business units, data is essential to spotting opportunities, managing risks, and quickly adjusting their offerings in the right direction. They need fast access to the data and the ability to combine it in various ways depending on their specific needs. Instead, for IT, data is a resource to be optimized by controlling access and reducing variability—essentially limiting the ways data can be recombined—to keep management costs low. This tension between the needs of IT and the business units doesn't serve either side well, nor does it benefit the organization as a whole.

A **hub-and-spoke model**, where IT manages data up to a certain level of consolidation (the hub) and then allows business units to freely integrate and use it (the spokes), somewhat alleviates the problem but does not solve it. Business units, in order to operate on the data independently, still need IT to make the data available according to its operational logic and in line with its objective of optimizing resource use.

Therefore, it's not enough to just use an **enclosure strategy** that divides responsibilities between IT and business units along the data value chain. Instead, a **recommoning strategy** should be applied that mobilizes everyone in the organization to take collective responsibility for data. As Jobe Bloom describes in his *theory of the three economies* (check the *Further reading* section for *ReCommoning: Transitioning Organisations - Jabe Bloom (2022)*, `https://youtu.be/ mGNEbDhT1Hw?si=SUYflz7whMiAffvG`), organizations need to create an intermediate space between central IT and business units, where both can collaborate on data management. This would help reconcile the differing goals of IT and the business units, aligning their efforts in data management to better serve the organization as a whole (*Figure 12.3*).

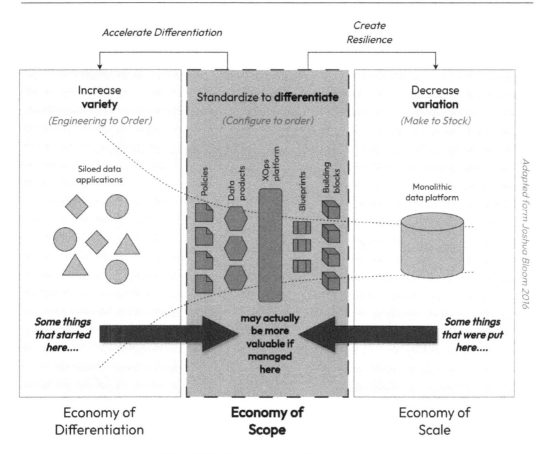

Figure 12.3 – Recommoning data management

For this reason, I believe it is essential to create an ecosystem within organizations where everyone, without distinction between producers and consumers, shares the responsibility for managing data and, more broadly, for managing the entire information architecture. In this book, I have described this ecosystem, outlining the rules for governing it (including organizational architecture and the operating model) as well as the strategies for creating, mobilizing, and scaling it within organizations.

Setting yourself up for success

A popular Italian song I love has a line that goes something like this: "People are known to give good advice when they can no longer set a bad example." While I hope to continue setting the "bad example" for a long time to come, I feel I cannot leave you without a few suggestions in this final chapter—hopefully, good ones.

Be optimistic, not naïve

Managing data as products is a paradigm shift that has both technological and organizational impacts. As we have emphasized multiple times throughout the book, it is important to ensure that there is a sufficient level of sponsorship and resources before starting. Moreover, during the adoption process, it is essential to work intensively to gain buy-in from the various groups that will be progressively involved.

Change processes are inherently complex. While maintaining optimism about the success of transformation initiatives is important, it is equally crucial to remain grounded in reality. Naivety must be avoided, and efforts should not be wasted on challenges that are unlikely to succeed.

At each stage of the change process, it is vital to prepare the groundwork for subsequent steps. This entails involving all relevant stakeholders to discuss ideas, gather input, and obtain the necessary sponsorship and buy-in before proceeding. In Japanese culture, this preliminary consensus-building is referred to as **nemawashi**, which translates to *"digging around the roots."* Effective change, akin to transplanting a plant, requires addressing the roots first and ensuring they are freed from their existing environment.

Thinking that change can happen solely from the bottom up, without support from above, or purely from the top down, relying only on imposition without grassroots buy-in, is a sure path to failure for the transformation initiative.

> *"The organizational change process is not top-down or bottom-up, but participative at all levels—aligned through a common understanding of the system"*
>
> —Peter Senge in The Fifth Discipline

Be a reflective practitioner, not a methodological purist

A methodology, including the one presented in this book, offers a conceptual framework for understanding a related class of problems and grounding the development of a practice to address them. However, each methodology, no matter how detailed, remains an abstraction built upon a conceptual model of the problems it aims to solve. Real-world issues are rarely identical to the model of the problem on which the methodology is based. In the real world, practice is rarely a straightforward and direct application of the methodology it was designed around. Therefore, it is essential to engage with any given methodology—such as managing data as a product—pragmatically, adapting it to the specific context in which it is applied.

It is also important to continuously reassess the practice derived from the adopted methodology, both during its execution (**reflection-in-action**) and in retrospect (**reflection-on-action**), to foster its ongoing improvement. The objective is to maximize the effectiveness of the practice by improving at each iteration its most critical elements—particularly those that have the greatest influence on the problems being addressed.

Being open to adapting a practice according to the context and continually refining it based on experiential learning are fundamental elements for successfully implementing a practice within an organization. It is vital to acknowledge that no new practice, regardless of the level of support and resources available, has an unlimited timeframe to demonstrate its value. Thus, reflection and action should merge into a lean approach to implementing and adopting the practice, making it perpetually open to adaptation and improvement.

> *"When someone reflects-in-action, he becomes a researcher in the practice context.*
> *He is not dependent on the categories of established theory and technique, but*
> *constructs a new theory of the unique case."*

> —*Donald Schön in The Reflective Practitioner*

Focus on the system, not on the parts

A data management solution is a complex socio-technical system. Building it requires a holistic approach, where each core component that ensures the system's viability—its ability to operate autonomously and adapt to its environment—must be developed and evolved in synergy.

Specifically, the four key capabilities that the data management function needs at an operational level—**data product development**, **governance policy-making**, **platforming**, and **enabling**—are highly interdependent, and thus it is counterproductive to implement only some while neglecting others. Depending on the context and the stage of the journey in adopting the managing data-as-a-product paradigm, one of these capabilities may require more attention, necessitating greater investment in its evolution. However, this evolution must occur in harmony with the others to avoid generating negative externalities that could create bottlenecks in other parts of the system.

Even as objectives, responsibilities, and resources are distributed across different parts of the system, it's crucial to maintain a clear view of the whole—focusing on how these parts interact. A complex system such as a data management solution thrives not when each part performs at its peak in isolation, but when they work together effectively to support the system's goals. Start each initiative by examining these interactions, identifying the next best step to optimize this intricate web of interdependencies.

> *"A system of local optimus is not an optimal system at all: it is a very*
> *inefficient system"*

> —*Eli Goldratt in The Goal*

Decentralize to scale, do not scale to decentralize

Modularization is essential for managing complexity. The best way to address a complex problem is to break it down into smaller, more manageable parts. This need to divide complex problems is a key driver behind the managing data-as-a-product paradigm, where data products act as the modular units in the data architecture described in this book.

In addition to simplifying complexity, modularizing a system enables the distribution of responsibilities across different parts, decentralizing them. This decentralization gives each part the autonomy to define its own development path, enhancing the system's agility in adapting to its environment. For a data management solution, this means the organization can more effectively use data to compete in its market.

However, there's no such thing as a free lunch. Decentralizing responsibilities increases agility but also adds complexity, especially in alignment and coordination. If this complexity isn't managed, agility can quickly turn into chaos. Managing coordination costs is far from marginal.

For this reason, it is essential to always assess the level of agility needed by the organization and adjust the degree of decentralization accordingly. Once the desired agility level—reflecting the system's **optimal complexity**—is achieved, further decentralization is unnecessary and only adds complications, where the cost of managing these complexities outweighs any extra benefits in agility (i.e., the **principle of satisficing**).

> *"Everything should be made as simple as possible, but no simpler"*
>
> —*Albert Einstein*

Be a change agent, not a change manager

Every change generates resistance. No matter how big the problem is, how urgent the need to solve it is, or how promising the proposed solution may seem, resistance to change will always be high. This resistance, of course, does not come from the technologies but from the people. Technologies do not have opinions, goals, fears, habits, or feelings—people do.

As we've noted before, in a sociotechnical system, the organizational and technological aspects must evolve together in a synergistic way. However, this doesn't mean that people and technology hold equal weight in the change process. We cannot treat the change process as purely mechanical, where people are merely resources to be reconfigured in order to transition to a new state. Instead, people should be at the center of change, actively participating rather than passively accepting it.

This highlights the importance of acting as a change agent rather than just a change manager. A change manager focuses on the process of change itself while a change agent focuses more on the people involved in it, seeking to influence, inspire, encourage, and reassure them.

A change agent must be a good listener. Most people do not resist change because they do not understand its benefits but because they fear what they will lose as a result. Therefore, it makes little sense to insist too much on the benefits of change when trying to engage someone actively. It is more important to listen to understand what people fear losing as a result of the change.

A change agent must bring these fears to light and recognize their importance, working to alleviate them. This task can be challenging and requires patience. Many people prefer to oppose change either actively and vocally (on-stage opponents) or passively and silently (off-stage opponents), often criticizing potential outcomes rather than openly sharing their personal fears. Sometimes, these fears—the root cause of resistance—aren't even conscious to those opposing changes. However, these fears must be brought to the surface and managed for change to proceed successfully.

> *"The major problems of our work are not so much technological as sociological in nature"*

> —*Tom DeMarco in Peopleware*

Be curious, not obsessed

Being passionate about data management and constantly seeking to improve one's knowledge in this field is key to success in the practice. However, data management is increasingly extending beyond just the technical aspects of handling data to encompass other disciplines, such as economics, sociology, semiotics, systems theory, cybernetics, and more. As a result, professionals must look beyond their specialization and cultivate a genuine curiosity that drives cross-disciplinary exploration. An open, interdisciplinary mindset enriches one's cultural and professional knowledge while providing the tools to tackle complex data challenges in innovative and creative ways.

The aim is to become a **T-shaped professional**—someone who combines deep expertise in a specific area with a broad understanding of other fields. The horizontal bar of the T symbolizes this openness, enabling effective communication with experts from different domains, grasping diverse perspectives, and collaborating within multidisciplinary teams. Being T-shaped means navigating your specialized area and related fields with agility, accepting that while you can't know everything, there's always an opportunity to learn.

> *"Don't be a know-it-all; be a learn-it-all"*

> —*Satya Nadella*

Final remarks

Before wrapping up, I want to share a couple of final thoughts to provide proper context for everything we've discussed so far regarding the data product-centered paradigm and data management in general.

The evolution of data management

Data management is inherently complex. The paradigm outlined in this book acknowledges this complexity and seeks to offer a data product-centered paradigm for navigating it more effectively. However, managing data as a product is not the ultimate solution to every challenge in data management. It is not a paradigm that renders all previous methods obsolete and research for new ones unnecessary. In other words, it's a significant step in the evolution of data management practice, not the final chapter.

Managing the data-as-a-product paradigm provides a fresh way to move forward, correcting past mistakes while also building on valuable lessons from past experiences. Therefore, adopting this paradigm doesn't mean we no longer need to understand previous paradigms, nor does it imply that we should stop researching and keeping up with new paradigms that will emerge in the future. Ultimately, our goal as practitioners is to navigate the evolving landscape of data management with a mindset that combines innovation with respect for the foundations that got us here.

Data-centered and people-driven organizations

Data, and more broadly information architecture, provides the context needed for organizations to make better decisions more quickly. The role of data and its management is strategic, as an organization's ability to compete in the market relies on the speed and quality of its decision-making. However, data doesn't make decisions on its own. It's used by people, and ultimately, it's people who make the decisions. The experience and wisdom of individuals are central to an organization's success. Even the best pilot in the world would struggle to fly an airplane blindfolded. However, the challenges for someone with no knowledge of flying would be even greater, no matter how advanced the flight instruments are. Therefore, while data is essential, it is not a guarantee of success by itself. People, supported by data, are what truly drive success. For this reason, organizations need to become data-centered while always remaining people-driven.

Summary

Here we are at the end of the end. The topics covered have been many, and I know that some chapters are particularly dense with content. I have deliberately tried to describe managing data as a product paradigm in an extensive and detailed manner, hoping that what I've written can assist you not only in terms of vision but also in practice. If you've made it this far, I'm glad and grateful for your attention and the time you've spent reading the book. I truly hope that the "AHA" moments have outnumbered the "WTF" moments.

If you have any questions, curiosities, doubts, or feedback, feel free to reach out to me directly on LinkedIn (`https://www.linkedin.com/in/andreagioia/`) or use the discussion forum on the GitHub page associated with the book (`https://github.com/PacktPublishing/Managing-Data-as-a-Product/`).

Now, the ball is in your court. It's time to pick it up and start playing with the passion and commitment needed to achieve your goals, all while remembering to enjoy the journey. Good luck!

Further reading

For more information on the topics covered in this chapter, please see the following resources:

- *Post-Capitalist Society* – P. Drucker (1994) `https://a.co/d/gaTZxSs`

- *Competitive Advantage: Creating and Sustaining Superior Performance* – M. E. Porter (1998) `https://a.co/d/4jyYYbZ`

- *Resource-Based Theory: Creating and Sustaining Competitive Advantage* – J. B. Barney, D. N. Clark (2007) `https://a.co/d/auNESHi`

- *Dynamic Capabilities and Strategic Management: Organizing for Innovation and Growth* – D.J Teece (2009) `https://a.co/d/bBtc8PD`

- *ReCommoning: Transitioning Organisations* – Jabe Bloom (2022) `https://youtu.be/mGNEbDhT1Hw?si=SUYflz7whMiAffvG`

- *The Fifth Discipline: The Art & Practice of The Learning Organization* – P. M. Senge (2006) `https://a.co/d/hvNUIyf`

- *Fearless Change: Patterns for Introducing New Ideas* – L. Rising, M. L. Manns (2004) `https://a.co/d/0GQUgcZ`

- *The Reflective Practitioner: How Professionals Think In Action* – D. A. Schon (1984) `https://a.co/d/etSQprf`

- *The Goal: A Process of Ongoing Improvement* – E. M. Goldratt, J. Cox (2012) `https://a.co/d/6GW4bCm`

- *Peopleware: Productive Projects and Teams* – T. DeMarco, T. Lister (2016) `https://a.co/d/1SvJIQZ`

Get This Book's PDF Version and Exclusive Extras

UNLOCK NOW

Scan the QR code (or go to `packtpub.com/unlock`). Search for this book by name, confirm the edition, and then follow the steps on the page.

Note: Keep your invoice handy. Purchases made directly from Packt don't require one.

13
Unlock Your Exclusive Benefits

Your copy of this book includes the following exclusive benefit:

- ☁ Next-gen Packt Reader
- 📄 DRM-free PDF/ePub downloads

Follow the guide below to unlock them. The process takes only a few minutes and needs to be completed once.

Unlock this Book's Free Benefits in 3 Easy Steps

Step 1

Keep your purchase invoice ready for *Step 3*. If you have a physical copy, scan it using your phone and save it as a PDF, JPG, or PNG.

For more help on finding your invoice, visit `https://www.packtpub.com/unlock-benefits/help`.

> **Note**
> If you bought this book directly from Packt, no invoice is required. After *Step 2*, you can access your exclusive content right away.

Step 2

Scan the QR code or go to `packtpub.com/unlock`.

On the page that opens (similar to *Figure 13.1* on desktop), search for this book by name and select the correct edition.

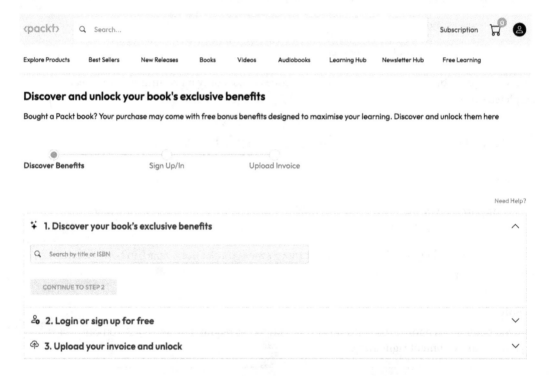

Figure 13.1: Packt unlock landing page on desktop

Step 3

After selecting your book, sign in to your Packt account or create one for free. Then upload your invoice (PDF, PNG, or JPG, up to 10 MB). Follow the on-screen instructions to finish the process.

Need help?

If you get stuck and need help, visit
`https://www.packtpub.com/unlock-benefits/help`
for a detailed FAQ on how to find your invoices and more. This QR code will take you to the help page.

Note

If you are still facing issues, reach out to `customercare@packt.com`.

Index

X

Packtpub.com

Subscribe to our online digital library for full access to over 7,000 books and videos, as well as industry leading tools to help you plan your personal development and advance your career. For more information, please visit our website.

Why subscribe?

- Spend less time learning and more time coding with practical eBooks and Videos from over 4,000 industry professionals

- Improve your learning with Skill Plans built especially for you

- Get a free eBook or video every month

- Fully searchable for easy access to vital information

- Copy and paste, print, and bookmark content

Did you know that Packt offers eBook versions of every book published, with PDF and ePub files available? You can upgrade to the eBook version at packtpub.com and as a print book customer, you are entitled to a discount on the eBook copy. Get in touch with us at customercare@packtpub.com for more details.

At www.packtpub.com, you can also read a collection of free technical articles, sign up for a range of free newsletters, and receive exclusive discounts and offers on Packt books and eBooks.

Other Books You May Enjoy

If you enjoyed this book, you may be interested in these other books by Packt:

Driving Data Quality with Data Contracts

Andrew Jones

ISBN: 978-1-83763-500-9

- Gain insights into the intricacies and shortcomings of today's data architectures
- Understand exactly how data contracts can solve prevalent data challenges
- Drive a fundamental transformation of your data culture by implementing data contracts
- Discover what goes into a data contract and why it's important
- Design a modern data architecture that leverages the power of data contracts
- Explore sample implementations to get practical knowledge of using data contracts
- Embrace best practices for the successful deployment of data contracts

Building Modern Data Applications Using Databricks Lakehouse

Will Girten

ISBN: 978-1-80512-333-0

- Deploy near-real-time data pipelines in Databricks using Delta Live Tables
- Orchestrate data pipelines using Databricks workflows
- Implement data validation policies and monitor/quarantine bad data
- Apply slowly changing dimensions (SCD), Type 1 and 2, data to lakehouse tables
- Secure data access across different groups and users using Unity Catalog
- Automate continuous data pipeline deployment by integrating Git with build tools such as Terraform and Databricks Asset Bundles

Packt is searching for authors like you

If you're interested in becoming an author for Packt, please visit authors.packtpub.com and apply today. We have worked with thousands of developers and tech professionals, just like you, to help them share their insight with the global tech community. You can make a general application, apply for a specific hot topic that we are recruiting an author for, or submit your own idea.

Share your thoughts

Now you've finished *Managing Data as a Product*, we'd love to hear your thoughts! Scan the QR code below to go straight to the Amazon review page for this book and share your feedback or leave a review on the site that you purchased it from.

https://packt.link/r/1-835-46853-5

Your review is important to us and the tech community and will help us make sure we're delivering excellent quality content.

www.ingramcontent.com/pod-product-compliance
Lightning Source LLC
Chambersburg PA
CBHW080613060326
40690CB00021B/4676